C000152863

The Worldwide Crisis in Fisheries
Economic Models and Human Behavior

The world's marine fisheries are in trouble, as a direct result of overfishing and the overcapacity of fishing fleets. Despite intensive management efforts, the problems still persist in many areas, resulting in many fisheries being neither sustainable nor profitable. Using bio-economic models of commercial fisheries, this book demonstrates that new management methods, based on individual or community catch quotas, are required to resolve the overfishing problem. Uncertainty about marine systems may be another factor contributing to overfishing. Methods of decision analysis and Bayesian inference are used to discuss risk management and the precautionary principle, arguing that extensive marine reserves may be the best way to protect fisheries, alongside a controlled catch quota system. This book will be of interest to environmental scientists, economists and fisheries managers, providing novel insights into many well-known but poorly understood aspects of fisheries management.

COLIN W. CLARK is Professor Emeritus in the Department of Mathematics at the University of British Columbia.

The Worldwide Crisis in Fisheries
Economic Models and Human Behavior

COLIN W. CLARK
University of British Columbia

Bob,
This was your idea!
With best regards,
Colin Clark
6 May, 2007

CAMBRIDGE UNIVERSITY PRESS

CAMBRIDGE UNIVERSITY PRESS
Cambridge, New York, Melbourne, Madrid, Cape Town, Singapore, São Paulo

Cambridge University Press
The Edinburgh Building, Cambridge CB2 2RU, UK

Published in the United States of America by Cambridge University Press, New York

www.cambridge.org
Information on this title: www.cambridge.org/9780521840057

© Cambridge University Press 2006

First published 2006

Printed in the United Kingdom at the University Press, Cambridge

A catalog record for this publication is available from the British Library

Library of Congress Cataloging in Publication data

ISBN-13   978-0-521-84005-7 hardback
ISBN-10   0-521-84005-8 hardback

ISBN-13   978-0-521-54939-4 paperback
ISBN-10   0-521-54939-6 paperback

Contents

Introduction

The current worldwide crisis in marine fisheries can be summarized as "too many boats chasing too few fish." In other words, many marine fisheries today suffer from a combination of overfishing of stocks and overcapacity of fishing fleets. Although fishing vessels can, and do, switch to new targets as current stocks become depleted, the world's oceans are finite. By the end of the 1990s the inevitable result was becoming apparent, as total worldwide catches of ocean fish (as reported by the United Nations Food and Agriculture Organization) began to decline. Documentation of this trend is provided by Pauly et al. (1998), Jackson et al. (2001), Hilborn et al. (2003), Myers and Worm (2003), Dulvy et al. (2003) and other publications. Based on large, detailed databases for world fisheries, these studies have demonstrated an unexpected degree of overharvesting, especially of large, valuable species.

Two important questions immediately arise. What are the underlying reasons for overfishing and overcapacity? What needs to be done to reverse the trend?

To begin with, it is obvious that economic forces are paramount— marine resources are exploited because of a demand for the product. If the revenue obtained from catching fish of a certain population exceeds the cost of doing so, there will exist an economic incentive to exploit that population. Unless the rate of harvesting can be controlled somehow, the fish population may eventually be reduced (at a profit) to a low level. This in turn may affect the productivity of the resource and greatly reduce future catches. In extreme cases, the population may become extinct (Dulvy et al. (2003) document 133 local, regional and global extinctions of marine populations), but in other cases the population may persist at a low level, characterized by the term "bionomic equilibrium."

Bionomic equilibrium is defined as the stock level at which revenues

1

and costs of fishing are equal. It is an economic equilibrium because fishing is profitable at higher, and unprofitable at lower, stock levels. Whether bionomic equilibrium implies low biological productivity thus depends on the cost–price ratio. The more valuable the species, and the lower the cost of catching it, the more heavily will it be exploited. Bionomic equilibrium then has a double defect—low catch rates and low (or zero) economic profitability. Both defects could seemingly be avoided if the fishery could be prevented from ever reaching bionomic equilibrium. This quest has been the basis of fishery management for the past 50 years or more. How is it implemented?

The traditional approach is to first determine some "optimal" level of annual harvests (or of fishing effort), and then to control fishing to achieve that objective. Surprisingly, perhaps, this approach has often failed to prevent overfishing. To quote Caddy and Seijo (2005), "This is more difficult than we thought!" The present book attempts to diagnose the difficulties of fisheries management, using analytic models (as opposed to computer simulation models) that explicitly incorporate the economic incentives of fishermen or vessel owners.

To briefly summarize the diagnosis, imagine that a TAC (Total Allowable annual Catch) program is introduced into a certain fishery that has previously been overfished. Initially the TAC is set low enough to allow the population to recover to some target level. (In practice a severely depleted population may recover very slowly, or fail to recover entirely, but we ignore this possibility here.) Once the recovery has occurred a sustainable TAC is used (but subject to annual adjustments necessitated by natural fluctuations in recruitment). What happens next?

Since the fish stock is now at a higher level than bionomic equilibrium, fishing has again become profitable. Therefore new fishermen will now be attracted to the fishery. The fishing fleet will expand. A new equilibrium, called "regulated bionomic equilibrium," will emerge when the costs of operating this expanded fleet again balance the revenues from fishing. Prediction: *TAC-based fisheries management inevitably leads to the overcapacity of fishing fleets*—unless steps are taken to prevent this expansion.

To repeat, much of today's excess fishing capacity in managed fisheries may be a direct economic consequence of the management system itself. (In many cases, subsidization of the fishery, though undertaken with the best of motives, has further exacerbated the problem of overcapacity.) Overcapacity adds to the difficulties of management, since fishermen require large catches in order to obtain enough income to meet their fixed costs. Any downwards adjustment to the TAC, required for conservation purposes, is strongly resisted by the fishing industry. Thus the management process degenerates into a struggle between managers and fishermen—even though the avowed purpose of management is to protect the long-term interests of the fishermen.

The next obvious step is to try to reduce excess capacity—but how? Fishermen will not willingly withdraw their vessels unless they can find other profitable fishing opportunities. The idea has therefore arisen to pay some of the fishermen to leave the fishery. A vessel buy-back program is introduced, whereby vessels are bought up (and presumably destroyed) by a benevolent government. This relieves the pressure on the fish population, and improves profitability for the remaining vessels. Then what?

Note that the buy-back program has renewed the economic incentives for fleet expansion. Clearly these incentives must be thwarted, by allowing only licensed vessels (e.g. those that remained after the buy-backs) to participate. Such "limited entry" management systems do not counter the incentives for expansion, however; they merely outlaw it. Experience with limited entry programs in many countries has been that the owners of licensed vessels tend gradually to increase the fishing power of their vessels in a process that is termed "capital stuffing." Examples of capital stuffing include increased engine horsepower, improved navigational systems, upgraded fishing gear, and increased freezer capacity. These subtle changes can, over time, substantially increase the total fleet's fishing capacity, necessitating a second round of buy-backs, and so on.

All of these difficulties can be traced to a single source, the common-pool nature of the fishery resource. The exploiters of any common-pool resource stock inevitably participate in a prisoners' dilemma type of game. Each fisherman attempts to maximize his (or her) personal gain from fishing, but the system ends up in an equilibrium situation where everyone gains nothing. Cooperative behavior, if it were possible, could achieve a situation in which everyone makes positive gains.

How can fisheries management be revised so as to produce a positive result? One method that is sometimes suggested is for government to tax resource users, for example by charging a royalty on all landed fish. In this way the government asserts its ownership of the resource, charging users accordingly. By imposing the appropriate tax rate, the government can in principle force the fishery to operate in an optimal, sustainable mode. The fish stock is driven to the "taxed bionomic equilibrium," where fishermen earn zero net after-tax gains. The fishery does provide positive economic gains to society (through government revenues), but none of these gains accrue to the fishermen. Tax-based management of commercial fisheries has seldom been attempted, at least not as the sole regulatory instrument.

An alternative approach, currently in use in several fishing nations, is based on individually allocated annual catch quotas, or IFQs (individual fishing quotas). Provided that the quotas are rigorously enforced (since quota-busting would be profitable), an IFQ system has the potential for breaking through the trap of overfishing and overcapacity. The incentives for capital stuffing are removed, because fishermen have nothing to gain from such behavior. By altering fishermen's economic incentives, IFQs tend to encourage resource conservation, which will enhance the quota owners' future incomes. Of course this presumes that these future benefits will actually be received by existing quota owners, and not spread over a wider set of fishermen.

This latter question, however, raises another critical issue. In an IFQ system the quota owners may become quite wealthy, especially if trading of quotas is permitted. Is it socially justifiable for the government to grant long-term (or permanent) privileges to a select few? Or should the government charge quota owners a fair royalty for these restricted rights? Such royalties are commonplace in other resource industries, and are probably destined to become so in national marine fisheries as well. IFQs can encourage sustainable fishing, and levying reasonable royalties or fees can achieve an equitable sharing of economic benefits.

Any management system relies on specifications of the appropriate total annual catch quota (TAC), which may then be divided among quota holders on an annual basis. Determining the TAC is by no means simple, however. Traditionally this has been done using mathematical (or computer) models of fish population dynamics, from which an estimate of the current TAC can be derived (Smith 1994). But this process is subject to extreme levels of uncertainty. The size of the fish stock cannot be observed directly, but only inferred from rather unreliable data. Details of population structure, distribution and dynamics are likewise poorly understood, as are environmental factors that may affect the population.

How should management strategy respond to such uncertainty? This topic is currently under intense discussion. Phrases such as "risk management" and "the precautionary principle" are often encountered but exactly what these imply in operational terms remains somewhat obscure. An additional difficulty pertains to the complex nature of marine ecosystems. Should fisheries models attempt to encompass the full ecosystem? What data would be needed to support ecosystem models? Should TACs be calculated using ecosystem models, or are traditional single-species models sufficient, if intuitive adjustments are made to allow for interspecific effects? Or is an entirely new approach needed, for example incorporating large-scale marine reserves as a hedge against uncertainty and management error?

Fishery management is now undergoing rapid changes, particularly in terms of introducing individual fishing quotas, and managing conservatively for both profitability and sustainability. It seems likely that in most countries, institutional changes will be required to ensure that management is based on solid scientific and economic principles, and that risks and uncertainties are fully taken into consideration.

This book attempts to address these complex issues through the use of relatively simple (but perhaps not overly simple) bioeconomic models. Such models, which will doubtlessly be elaborated upon by others, are helpful in trying to understand the difficult problems of managing marine fisheries for the long-term benefit of producers, consumers, and resource owners—you and I. There is no single panacea for managing our marine fisheries, but a combination of the techniques analyzed in this book holds promise for resolving the present crisis of overfishing and overcapacity.

Acknowledgements

I am greatly indebted to John Beddington, Gordon Munro and Marc Mangel for invaluable comments and suggestions. Janet Clark expertly set the book in LaTeX.

1

Perspective

The proper study of mankind is man, said Pope. Yet our understanding of human behavior remains imperfect, if not rudimentary.

One theoretical development of the mid-twentieth century that has helped to improve our understanding of behavior is the theory of games (von Neumann and Morgenstern 1947; Nash 1951, 1953), which is now widely used in economics, and also in evolutionary biology (Maynard Smith 1982). Although I do not make extensive direct use of game theory in this book, I do emphasize that the economically motivated behavior of fishermen must be considered as an integral component of any fishery system.

Unfortunately, the economic theory of commercial fisheries is still widely misunderstood. Many fishery management programs have failed dismally to achieve their objective of conserving fish stocks. Although the specific reasons for a given failure are seldom apparent, it seems clear that in many cases the failure was *fully predictable* on basic economic principles.

To put the matter succinctly, many management programs have not attempted to deal directly with the economic motives that always underlie overfishing. Indeed, many programs may have inadvertently increased the motivation for overfishing. In addition, these programs have often encouraged the development of excess fishing capacity, now recognized as a major problem in world fisheries.

Let me be more explicit. First, it is self-evident that an unregulated open-access renewable resource stock will, if it is marketable, be exploited and perhaps depleted. Simply stated, if money can be made by catching fish in a certain area, those fish will sooner or later be caught. Eventually the stock of fish will become smaller (either fewer fish, or smaller individuals, or both), and fishing may become unprofitable. At this stage, various things can happen. For example, the fish stock may continue to produce a sustainable annual catch. This situation has been called the "bionomic equilibrium" of the open-access (or common-property) fishery (Gordon 1954). How large the sustained catch will be depends on a number of circumstances, including the biology of the fish population and the economic conditions of the fishery. Generally speaking, the more valuable the fish are, and the less costly it is to catch them, the *lower* will be the sustained yield. Thus, valued species will tend to be overfished, unless specific actions are undertaken to prevent this from happening. The history of global marine fisheries over the past 100 years (or so) strongly reflects this phenomenon of overfishing of highly valued species (Pauly et al. 1998; Myers and Worm 2003).

In some circumstances, until recently considered to be unusual, the fish population may virtually disappear and fishing cease. Even if the population is not actually biologically extinct (e.g., a remnant population may persist in some remote location), this situation is sometimes referred to as "economic" extinction.

Overfishing is generally considered to be undesirable. If annual catches had been controlled, it should have been possible to achieve a substantial harvest over the long run. Most existing fisheries management agencies have been established with the purpose of preventing overfishing, thereby maintaining a substantial sustainable yield. All participants should benefit from such a program—the fishermen will have a perpetual source of income, and consumers a perpetual supply of fish.

In practice, it has often proved to be amazingly difficult to accomplish this result. Why should that be so?

In fact, many diverse factors may be implicated in the persistent phenomenon of overfishing of managed fisheries. Examples of such factors include unanticipated environmental changes, erroneous models of fish population dynamics, erroneous estimates of stock abundance, unrecorded or illegal catches, and so on. It is my contention, however, that *normal economic behavior* by fishermen is almost always at the root of the overfishing problem.

To explain this assertion, it is necessary first to realize that in order to provide a substantial, steady annual yield, a given fish population must be maintained at a fairly high level, typically at least 50–60% of the natural, unfished population level. But any such population is highly attractive to the fishermen. Removing fish from such a healthy population may be highly profitable. (If not, the overfishing problem wouldn't exist.) *Overfishing is usually economically profitable* in the short run, except for fish species that are not marketable, or only marginally so.

But surely, one might argue, the fishermen understand that their long-term well-being depends on a healthy fish population, and they will not wish to act so as to deplete the stock. This, I maintain, is not necessarily the case, for several reasons.

First and foremost is the following game-theoretic aspect: *What is economically desirable for fishermen A depends on what the other fishermen are doing.* Fisherman A might act to limit his catches, to help preserve the stock, but this will be of no use to A if fishermen B, C, D keep on taking large catches. Worse, even if fishermen A, B, C and D agree to limit their catches, this may serve only to attract new fishermen E, F, ...

How can these economically motivated activities of competing fishermen be counteracted so as to maintain a sustainable and profitable fishery? Each of the following approaches has been seriously recommended at one time or another:

1. Impose a tax on all fish caught, thereby reducing or eliminating the economic incentive for overfishing.
2. Reduce fishing pressure by closing the fishery each year, as soon as the annual "allowable catch" has been taken.
3. Privatize the fishery by granting exclusive ownership to a single firm, which will then automatically have an incentive to maintain stewardship of the resource.
4. License a fixed number of fishing vessels calculated so as to be capable of capturing the specified annual allowable catch.
5. Award annual individual fishing quotas to each of a specific set of fishermen, the sum of the quotas being equal to the annual allowable catch.
6. Form a fishermen's cooperative, which will share the annual catch equally among its members (all other potential fishermen being excluded). Such a cooperative will then have an automatic incentive favoring stewardship of the resource.
7. Establish marine reserves, thereby ensuring the survival of at least part of the populations of depleted species.

At first glance it might appear that any one of these methods should work quite well. More careful consideration, however, reveals that there could be major differences between the approaches. Some methods may be more effective than others. How can one determine the best management approach? Or at least, an approach that results in sustainable, profitable fishing?

Several of the methods listed above have in fact been used in fishery management, sometimes with quite unexpected and undesired consequences. In some cases, the expected results of a given method were deduced from a simple model, which in fact turned out to be strongly misleading. As mentioned earlier, few of the models actually used in fishery management have considered the economic behavior of fishermen at all, and those that have done so have often been proved to be inadequate. The main purpose of this book is to identify such model failures, and to propose improved models.

One thing is immediately evident: all seven methods assume the existence of a central government that claims jurisdiction, and is prepared to exert control, over the fishery resource. Without this jurisdiction and control, methods of preventing overfishing are limited, although international treaties have sometimes been effective. Such treaties are particularly useful in the case of transboundary resources shared by two or more coastal states.

It is worthwhile here to pause briefly to discuss basic terminology, which has sometimes been used carelessly. First, I use the phrase "unregulated open-access fishery" to refer to a fishery in which anyone may participate without any form of restriction or regulation. This was the situation assumed in the above discussion of overfishing. Uncontrolled high-seas fisheries are now the main example of unregulated open access.

The term "common-property fishery" which has often been used as a synonym for unregulated open-access fishery suffers from imprecision, and will not be used in this book.

Other terminology will be introduced as required. Many current management programs, for example, can be categorized as "regulated open-access fisheries." This term describes a fishery with no limits on participation (except perhaps regarding nationality), but subject to a variety of regulations, such as total annual catch quotas (TACs), gear and vessel restrictions, closed seasons or areas, and so on. Models pertaining to these and other management techniques will be featured throughout the book.

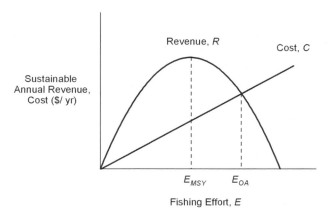

Figure 1.1 The Gordon model of fishery bioeconomics.

1.1 A Simple Bioeconomic Model

Figure 1.1 is a simple, well-known model of the unregulated open-access fishery (Gordon 1954). Understanding the advantages and the limitations of this model is a prerequisite for working with more detailed models later.

The horizontal axis in Fig. 1.1 represents fishing effort, E, defined here as an index of economic inputs (e.g., capital and labor) devoted to the fishery, on an annual basis. For example, in a trawl fishery, annual effort E is typically specified as the total number of standardized trawls per year, although shorter time periods may sometimes be used. Other specifications of effort are used in other cases. The annual costs of fishing C are assumed to be proportional to effort E:

$$C = cE \qquad (1.1)$$

Sustainable total annual revenue R is assumed to be given by

$$R = pY(E) \qquad (1.2)$$

where $Y(E)$ denotes the sustainable annual yield (catch), and p is the dockside price of fish. The sustainable yield curve $Y(E)$, and hence the sustainable revenue curve, are assumed for the moment to have an inverted U-shape, as shown. (The biological basis of the yield curve $Y(E)$ will be discussed in Chapter 2.) In particular, the yield curve $Y(E)$ achieves a maximum value $Y_{\max}$ at an intermediate level of effort $E = E_{\text{MSY}}$. This is the famous Maximum Sustained Yield (MSY) situation, long considered to be the ideal target of fisheries management.

The MSY objective, however, has been criticized on many grounds (see Larkin 1977, Mangel et al. 2002). For the moment let me just suggest that there may be a lot more to fishery management than simply maximizing the catch. For one thing, fish populations are dynamic systems that may respond gradually (or in some cases, rapidly) to human intervention. These changes can have important economic consequences, as we shall see in later chapters.

1.2 Bionomic Equilibrium

The first prediction of our present model, Fig. 1.1, is that in an unregulated, open-access fishery, effort will achieve an equilibrium E_{OA} at the level where annual revenues exactly match annual cost

$$pY(E_{\text{OA}}) = cE_{\text{OA}} \qquad (1.3)$$

This effort level E_{OA} is called the bionomic equilibrium effort level of the unregulated open-access fishery.

Bionomic equilibrium is a powerful and testable prediction, with profound management consequences. As we will also see, however, the prediction can be seriously misunderstood. (Macinko and Bromley, 2002, call this the "most misunderstood model in the (fishery) world.")

First, let me explain the prediction of bionomic equilibrium. Imagine a level of effort E below E_{OA}. Then (see Fig. 1.1) annual revenues exceed annual costs: fishermen are making money. Other potential fishermen will then be motivated to enter the fishery, and effort will tend to increase. This economic incentive for expansion persists as long as $E < E_{\text{OA}}$. On the other hand, E will not exceed E_{OA} over the long run, because economic incentives would then favor quitting the fishery. Thus $E = E_{\text{OA}}$ is a stable equilibrium for the unregulated open-access fishery.

In Figure 1.1, the bionomic level of effort E_{OA} is shown as being larger than E_{MSY}. In general, E_{OA} may be smaller or larger than E_{MSY}, depending on the relative cost and price parameters c and p. For example, a low-value, high-cost fishery will be lightly exploited, or even unexploited. Conversely, a high-value, low-cost fishery will be heavily exploited, and possibly overexploited. These common-sense predictions are a useful feature of the Gordon model.

Specifically, if the price-cost ratio p/c is high, then $E_{\mathrm{OA}} > E_{\mathrm{MSY}}$ (as in Fig 1.1). In this case the result of bionomic equilibrium is termed biological overfishing, because the annual yield $Y(E_{\mathrm{OA}})$ is lower than the MSY level, as a result of the depletion of the fish population. Precisely this situation is the principal feature of today's worldwide crisis in marine fisheries.

Now, how, exactly, is the Gordon model misunderstood? This question will be addressed throughout Chapters 1 and 2. (Then why bother studying the Gordon model, you may ask. Two answers are, first, that the important prediction of bionomic equilibrium is easily understood on the basis of the Gordon model. Second, "fixing" the Gordon model will be instructive in trying to understand the principles of fishery management—and the reasons underlying mismanagement.)

In interpreting the Gordon model, the term "cost" may be taken too simplistically. Here, as in economics generally, cost means opportunity cost, which is the cost involved in not engaging in one's most profitable alternative activity. The unit cost of effort, c, would thus include the costs of operating vessels—fuel, supplies, etc.—plus opportunity wages of captain and crew. These wages are not necessarily what fishermen actually receive, but their best wages from available alternative employment. Note that this concept of opportunity cost is central to the above argument for bionomic equilibrium. Fishermen will enter or leave a given fishery after comparing their expected net revenues with income opportunities (including fishing) elsewhere.

At bionomic equilibrium, therefore, the fishermen are earning "normal" incomes from the fishery—incomes that are at least as good as they could earn elsewhere. However, net sustained revenues are lower at bionomic equilibrium, possibly much lower, than could be achieved at some lower level of fishing effort. For example, net sustained revenues $R-C$ would be maximized at an effort level E_{MEY} ("maximum economic yield"), as depicted in Fig. 1.2. The difference $R - C$ between revenues and costs in a resource-based industry is formally called *economic rent*:

$$\text{Economic rent} = R - C \qquad (1.4)$$

Bionomic equilibrium of the unregulated open-access fishery is thus characterized by the complete *dissipation* of economic rent. Assuming that things could be otherwise, this dissipated rent constitutes a loss of wealth to society at large (Christy and Scott 1965).

It would be an improvement if this rent (or a major part of it) could

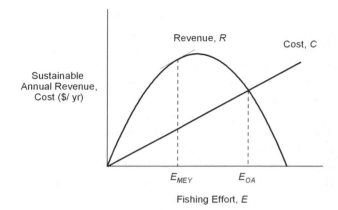

Figure 1.2 Maximum economic yield in the static Gordon model.

be preserved. But *who should benefit from the preserved rent?* Strange
to say, this question has been widely ignored in the existing literature
on fishery management and economics, an exception being an important study entitled "Who owns America's fisheries?" by Macinko and
Bromley (2002). I will take up this question in Chapter 4.

Positive economic rents are what motivate expansion of the fishery.
The management implications of this basic fact have often been misunderstood, or ignored. More on this in a moment.

Historically, both price p and cost c may vary over time, usually with
price increasing and cost decreasing over the years. What is especially
perverse about the open-access fishery is that price increases and/or
cost decreases ultimately worsen the performance of the fishing industry,
resulting in progressively more severe overfishing. Temporary gains from
price increases or cost decreases do occur, and it is these gains that drive
the expansion that eventually eliminates sustainable gains.

The dissipation of economic rent at bionomic equilibrium is often referred to as "economic overfishing." As noted earlier, bionomic equilibrium also often involves biological overfishing, in the sense that the fish
stock has become reduced to a level below that which provides the maximum sustained yield. Note, however, that the Gordon model of Figs. 1.1
and 1.2 makes no explicit reference to the stock level. Many misunder-
standings of the Gordon model are related to this lack of consideration
of the underlying resource stock, i.e., the fish population itself. This important limitation of the Gordon theory will be discussed in Section 1.7,
and in greater detail in Chapter 2.

The question arises whether bionomic overfishing can eventually destroy the fish population. This question cannot be decided on the basis of the present simple model. As explained in Chapter 2, bionomic equilibrium in an uncontrolled fishery can indeed result in the collapse of the fish stock, as has in fact often occurred (Pauly 1995; Myers et al. 1997; Mullon et al. 2005). The ultimate cause of such collapses is often controversial, but it is clear that reducing a fish population well below the MSY level is a dangerous act that can have unpredictable long-term consequences. Whether a severely overfished population will recover if fishing pressure is reduced or eliminated, is usually unpredictable (Hutchings 2000). These important questions are addressed frequently throughout the book.

1.3 Regulation of Fishing Effort

Referring again to Fig. 1.2, it is tempting to conclude that a simple solution exists to the overfishing problem. Fishing effort should be limited to some level below bionomic equilibrium. For example, the effort level E_{MEY} (for "maximum economic yield") would maximize sustained economic rents. This was in fact the original recommendation of Gordon (1954); it is still often encountered in the literature.

Several severe problems are associated with the prescription of MEY, however. Here I will concentrate on the two following difficulties:

(1) Almost without exception the fishing industry has displayed bitter opposition to the reduction of effort to achieve MEY.

(2) Management programs that have actually succeeded in reducing effort and increasing yields, have almost invariably been accompanied by an expansion of fishing capacity.

Both of these outcomes may seem counterintuitive. The problem is therefore to explain them, in terms of economic motivations. We begin with the fishermen's opposition to MEY—why should this happen? According to our model, a reduction in effort from bionomic equilibrium should produce two benefits, increased catches and revenues, and decreased costs of fishing. As several authors have proposed, this is surely a no-loser situation. So what's going on?

Let me begin by saying that part of the explanation is that our "stick-figure" model of Fig. 1.2 is highly oversimplistic at this stage.

To be specific, let us suppose that a certain fishery is currently at bionomic equilibrium. A reduction of effort, say by 50%, will definitely reduce costs of fishing. But will it really also increase yield and revenue? Usually not, in the short run. In most cases, reducing fishing effort will result in an approximately proportional *immediate* reduction in fish landings. Increased catches will occur only after the fish population has had a chance to recover from past overfishing. The recovery phase may be protracted—as long as decades or even centuries, in some cases. This fact is doubtlessly recognized by the fishermen, and was pointed out long ago by biologists (e.g. Beverton and Holt 1957; see also Berkeley et al. 2004).

Thus we see that overfishing may be reversible, but often only after a protracted recovery period. The more substantial the recovery the lengthier the recovery phase. Continued fishing during the recovery, even at a reduced effort level, will further lengthen the recovery period. The question then becomes, should the fishermen be willing to accept temporarily reduced, or zero, income to allow the fish population to regrow? For how long? Should the government pay the fishermen not to catch fish? Could the problem have been avoided if the use of excessive effort had been prevented to begin with? We will address such questions in Chapter 2, using a dynamic model. But at least we now have some idea why fishermen may not gladly accept a sharp reduction in effort.

This is not yet the whole story, however, by any means. In a sense, this entire book is the whole story, or at least a good part of it. Here I discuss another powerful effect (often ignored in economic analyses) that can cause fishermen to oppose a reduction in effort levels.

Imagine a fishery at bionomic equilibrium, but suppose now for simplicity that stock recovery would be immediate, following a reduction in effort. Indeed, suppose that biological overfishing does not occur, for example because annual recruitment to the fishery is exogenous. The analog to Fig. 1.1 then looks as shown in Fig. 1.3. Yield Y asymptotes at Y_{max} as effort E becomes indefinitely large. Bionomic equilibrium E_{OA} and maximum-rent effort E_{MEY} are specified as before.

An example is provided by the Florida spiny lobster fishery (Milon et al. 1999), in which annual recruitment is not linked to catch levels in previous years (recruitment of young lobsters is determined by the entire Caribbean lobster population, not by the relatively small sub-population exploited in the Florida fishery).

Any reduction in effort over a given year would have an immediate effect as indicated by Fig. 1.3. What happens in this fishery is that the

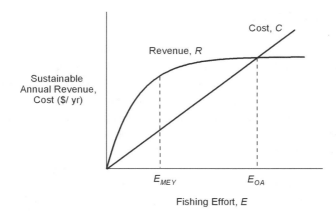

Figure 1.3 Bioeconomic equilibrium diagram for a saturating yield-effort curve $Y(E)$.

bionomic equilibrium effort level (approximately 540,000 lobster traps) captures over 90% of the annual crop of lobsters. Catch rates per trap decline over the season, as the lobster population becomes fished out. Reducing effort, by reducing the number of lobster traps, thus only reduces the annual catch by a relatively small fraction, unless the effort reduction is substantial. Nevertheless, lobster fishermen have expressed strong resistance to the degree of effort reduction proposed by the managing authority. What reason can be adduced for this situation?

The apparent answer is provided by Fig. 1.3 itself. Imagine that the model is more or less correct, so that reducing trap numbers by 50% (economists have recommended a reduction of about 65%) would lead to significant increases in annual catches per trap. The slight overall decrease in the total annual catches would be more than compensated by the 50% reduction in fishing costs, leading to increased profits for all fishermen—provided that ...

What extra proviso is needed here? In this fishery a trap certificate program (TCP) was initiated in 1993–4, with an issue of 747,000 certificates. By the 1998–9 season this number had been reduced to 544,000, close to the estimated value of $E_{OA} = 540,000$ traps. Why did the fishermen then oppose any further reduction in trap certificates? Economists had estimated that profitability would be maximized at between 160,000 and 200,000 traps.

The answer to this conundrum seems to be an increase in the level of cheating, actual or potential. Already by the early 2000s there were growing indications of illegal lobster fishing, through the use of unlicensed traps, stealing from licensed traps, lobsters taken illegally by divers, etc. This kind of activity is just what the bioeconomic model would predict— when profits turn positive, additional effort is attracted to the fishery. Such effort may be illegal, but enforcement of the regulations can be difficult, especially in this case, with over 500,000 trap certificates and a fishing area of over 100,000 square miles bordering the Florida Keys.

To repeat: management measures that reduce fishing effort from its bionomic equilibrium level will, in principle, result in increased rents for each existing fisherman. These increased rents may arise partly from cost savings, and partly (if gradually) from recovery of the fish population. But positive rents attract new effort. Unless this increase in effort can be prevented, the expansion will again result in the dissipation of economic rent.

Effort expansion in a managed fishery can occur in various ways, including illegal fishing, and increased fishing intensity by existing participants (see below). Rigorous enforcement is critical to the success of any management program. However, fishermen typically resist these efforts at controlling their activities. Economic incentives continue to operate in the managed fishery, because the fishermen continue to compete for the annual catch.

The seemingly irrational behavior of fishermen can also be regarded in game-theoretic terms. Nash (1951) argued that the players in a non-cooperative, non-zero-sum game would tend to reach a situation, nowadays called the Nash competitive equilibrium, in which no individual player can increase his gain by deviating from the equilibrium strategy. An unregulated open-access fishery fits this situation exactly. Prior to reaching bionomic equilibrium, economic rents are positive. The Nash competitive strategy is for each fisherman to catch as many fish as he profitably can. By failing to do so, an individual fisherman would obtain a smaller income, because other fishermen will catch the fish anyway. This is the same argument given earlier for bionomic equilibrium. The dynamics of the Nash competitive game lead to a combined biological and economic equilibrium in which economic rents are reduced to zero— the bionomic equilibrium.

The non-cooperative game model also applies to a regulated fishery, if the fishermen still compete for the total annual catch, and this is why so many management programs have failed. Section 1.6 below discusses an alternative approach, allocated quotas, that has the potential to transform fishing into a cooperative game. But first we discuss other important sources of management failure.

(The non-cooperative game that underlies bionomic equilibrium in fisheries is an instance of the ubiquitous "prisoners' dilemma." Two delightful books that emphasize the importance of prisoners' dilemmas in human and animal behavior are Barrett 2003 and Ridley 1998.)

1.4 Overcapacity

As mentioned above, overcapacity of fishing fleets has become a common occurrence worldwide, especially for managed fisheries. Generally speaking, the explanation for this phenomenon is again the fact that positive rents attract new effort. We will consider two different situations. First we imagine a newly developed fishery with no regulation whatever. For example, the fish population may have just been discovered, or a new market opportunity may have recently been recognized. Second we consider a mature fishery at bionomic equilibrium, into which a management program is then introduced.

In the first case, the new fishery begins with a large unexploited stock of fish. Catch rates are initially high, and many vessels are attracted to the fishery. Eventually the original stock is fished down, and catch rates decline sharply. The fleet that entered the fishery initially may be much larger than needed to catch the annual yield from the reduced stock. In this case overcapacity has been the result of a highly profitable initial phase of stock reduction. Unless some of the vessels can find other fishing opportunities, their best option may be to continue in the same fishery, resulting in excess fishing capacity. (In cases of extreme overfishing the vessel owner's best option may be to scrap the vessel and take up some other occupation. For example, when Antarctic whale stocks were fished out in the 1960s–1980s most European whaling fleets were either sold for scrap, or sold to Japan for refitting as deep-sea trawlers; see Clark and Lamberson 1982).

Next we turn to the case of a managed fishery, but one in which entry remains unlimited. Specifically we assume that the managers specify a total allowable annual catch quota (or TAC). Once the catch has been taken the fishery is then closed for the remainder of the year. If positive rents arise, what do we expect to transpire next?

This situation will be modeled in Chapter 2; here we provide an intuitive discussion. Since rents are positive (revenues exceed costs), existing fishermen are motivated to increase the catching power of their vessels. Also, additional fishermen are attracted to the fishery. Both of these events would tend to increase effort, but the managers prevent this by shortening the season. Are profits still positive, assuming that the TAC is maintained? Not necessarily, because the costs of fishing have now increased. These are primarily fixed costs (vessels and gear), which are not covered by our simple bioeconomic model. Nevertheless, the eventual outcome is predictable: fishing capacity will continue to grow until rents are again reduced to zero, at least approximately. A new, regulated bionomic equilibrium will eventually emerge, characterized by excess fishing capacity rather than by excess effort. In addition, the annual fishing season may have become quite brief.

Examples of regulated bionomic equilibrium abound. Let me briefly mention the Pacific halibut fishery, which has been managed, since the 1930s, by the International Pacific Halibut Commission, using area-specific TACs. Halibut is a popular table delicacy, and demand and price are high. By the late 1980s, fishing capacity had become so large that the halibut season was restricted to three days (two of which were partial fishing days) per year, in most areas. Since halibut can be taken for at least 180 days per year, this suggests that fleet capacity probably exceeded 60 times the level required to catch the TAC. Though extreme, this example is not atypical.

In 1991 the Canadian component of the halibut fishery was put under an ITQ (individual transferable quota) system. The wasteful and dangerous scramble of the pre-ITQ days has now largely disappeared. Also, fresh halibut (preferred to frozen halibut) has become available for much of the year in Canada. Pacific halibut fisheries in the USA were placed under ITQs beginning in 1995. However, processing companies complained that the ITQ system was damaging to them, as it required them to pay higher prices for the fish (Matulich and Clark 2003). The economics of ITQs are discussed in Chapter 4; an excellent review occurs in Ocean Studies Board (1999). But let me warn the reader that there are major problems associated with ITQs, particularly concerning who should receive the rents. See Section 1.6 below.

1.5 Subsidies

The rent-dissipation argument predicts that fishermen will often earn minimal incomes from fishing, even in fisheries that are successfully managed from a biological standpoint. This will be especially true in the case that fishermen have limited alternative income opportunities. Governments have therefore often undertaken to subsidize the fishing industry. Types of subsidies include vessel construction assistance, low-interest loans, price supports, seasonal unemployment benefits, and so on. The scale of fishing subsidies worldwide is large; for example, Munro and Sumaila (2002) estimated total subsidies for North Atlantic fisheries to be at least U.S. $2.5 billion per annum at the turn of the century.

The effect of fishing subsidies can readily be predicted. Since subsidies lower the cost of fishing, or increase revenues, or both, our analysis predicts that subsidies will lead to higher effort levels and greater depletion of fish stocks. If effort is regulated and depletion prevented, the result will be an expansion of fishing capacity and a progressive shortening of the fishing season. These are common features of today's marine fisheries, certainly in developed countries that possess the means for generous subsidization.

Buy-Back Programs

In recent years a new form of subsidy has become popular: buy-back subsidies. These are used to buy up excess fishing capacity, thereby, it is hoped, reducing fishing pressure on vulnerable stocks. There are two serious problems with buy-backs, however.

First, buy-back programs seldom achieve their objectives, for two reasons (Holland et al. 1999). The actual buy-backs are biased towards inefficient vessels and fishermen, leaving a residue of the more efficient operations. Thus the impact on overall fleet fishing power tends to be minimal. More importantly, if a buy-back program does succeed in temporarily reducing fishing capacity, this automatically renews the economic incentive for a new round of increases in capacity. Holland et al. (1999) describe several examples of this vicious cycle.

A second difficulty with buy-backs is that they may often be anticipated by the fishing industry. As shown in Chapter 3 (see also Clark et al. 2005), such anticipated buy-backs are economically equivalent to direct vessel subsidies. Therefore, such buy-backs will have the effect of motivating additional *expansion* of fleet size. Thus the billions of dollars currently being spent to buy back excess fishing capacity may have little effect on overfishing, and may even make the situation worse than before. (Not surprisingly, buy-back programs have been very popular with fishermen.)

1.6 Individual Quotas

The arguments presented so far may seem entirely negative, and indeed much of the current crisis in marine fisheries is the result of ineffective management schemes—TACs, limited entry, subsidies, buy-backs and other methods. Recently, however, a new approach has begun to be used, in which each fishing unit (fisherman, vessel, or enterprise) is issued an annual catch quota, or quota share. The sum of all the quotas equals the TAC. Suppose also that the individual quotas are transferable, in whole or in part, between owners. Each vessel owner will then be faced with the decision of using or selling his quota, or buying up additional quota units. Some owners will value quota units more highly than others. Thus economic considerations will motivate some owners to sell their quotas and leave the fishery. Those who continue in the fishery will be motivated to maximize the efficiency of their operations. Under ideal circumstances the fishery will tend to be exploited in an economically optimal manner.

Is this really correct? Can individual transferable quotas (ITQs) really lead to economically optimal results, without further controls? First, it is obviously essential that the quotas be rigorously enforced. Certainly there will be an incentive for quota holders to exceed their quotas, and also for poachers to take fish illegally. On the other hand, the quotas do constitute valuable assets, which their owners will wish to protect. A quota-owner may be particularly upset if others exceed their quotas, or if outsiders fish illegally. Thus peer pressure could help to ensure compliance with the quota system.

If ITQs are introduced into a fishery suffering from overcapacity, each individual quota may be too small to provide acceptable incomes for fishermen. Through consolidation of quotas, some fishermen will be able to operate profitably, while others leave the fishery. The market price for quota transfers will reflect the net value that a quota provides its owner over its tenure. Chapter 4 discusses bionomic models that explain how an ITQ system is likely to work.

ITQ systems can also be analyzed in game-theoretic terms, in this case using the theory of cooperative games. Quota owners will recognize that the value of their quotas (whether retained or eventually sold) depends on the future state of the fish population. Thus quota owners will be motivated to cooperate to ensure resource conservation—quite the opposite motivation that affects fishermen in the competitive situation with no individual quotas. We discuss these ideas in much greater detail in Chapter 4. Case studies involving cooperative behavior under ITQ systems are discussed in Chapter 6.

Difficulties Associated with ITQs

ITQ-based management also presents a number of difficulties. The importance of strict monitoring and enforcement has already been stressed. Another potential problem is the possibility of high-grading of catches through discarding of small-sized fish, thereby filling one's quota with high-value fish. This process can result in higher actual fishing mortality than planned for in setting the TAC. To make matters worse, the discarded catch will usually not be reported, and this bias can be a source of scientific uncertainty. Also, discarded by-catches of non-target species can result in overfishing of these populations. Many existing ITQ-based fisheries now use on-board observers (usually paid by the vessel owners) to help ensure accurate reporting of total catches, including discards. Chapter 6 describes several examples.

Another important and often neglected aspect of ITQs concerns social equity. A well-designed and well-run ITQ system can result in a highly profitable fishery. Consequently the individual quotas may take on very high monetary values. For example, Hilborn et al. (2005) report that "An Australian tuna fisherman recently sold his ITQ for southern bluefin tuna for $AUS 70 million."

Given that the purpose of an ITQ system is to prevent the dissipation of economic rent, or in other words to generate positive long-term benefits from the fishery, the enrichment of quota owners may seem inevitable. However, this immediately raises the question of fairness— why should the government initiate a system that establishes exclusive wealth-generating privileges for a chosen few?

Failure to face up to this question has led to distortion in the establishment of ITQ systems. For example, for many years the US Congress outlawed ITQs in American fisheries, basically because of the equity issue. Although this ban has now been removed, ITQs are still only slowly being implemented. Deciding on who gets the initial quota is difficult, and may be subject to challenges in the courts.

Another commonly encountered problem with ITQs is the concentration of quota ownership. In many cases the initial quotas have been quickly bought up by a few individuals, who may not themselves be active fishermen. The quotas are then leased back to fishermen, for example, by auction. Economic rents then accrue almost entirely to the new quotas owners (or to the original quota owners who sold off their

quotas for a quick profit). Minimal or zero rents now go to the active fishermen. This arrangement removes conservation-oriented incentives from
the fishermen. Many of the potential advantages of ITQs are weakened,
or disappear, under these circumstances, since the fishermen themselves
again have little to gain from conserving the resource.

To prevent concentration of quota ownership, some ITQ systems require that quota owners are active fishermen. While this regulation may
be useful, it can be difficult to enforce. It also fails to deal with another
problem, namely the difficulty that young people have in entering the
fishery, because of the high cost of buying a quota.

All of these problems result from the very success of the ITQ system
in terms of preserving economic rents, *assuming that the entire rents,
present and future, are received by the quota owners.* In this case the
initial recipients of the ITQs receive a windfall gain, namely the present
value of future rents from the fishery for the duration of the quotas. The
example of the Australian tuna fisherman indicates that the windfall
gain can be very large indeed.

The position advocated in this book is that such fisheries resource
give-aways always are distributionally and economically undesirable.
The general public is the ultimate owner of the marine resources within
its 200-mile zones, and deserves to receive a fair share of the rents derived
from those resources (see Macinko and Bromley 2002). Granting 100%
(or nearly 100%) of those rents to private individuals is unwarranted,
unwise, and unnecessary.

Two methods for sharing resource rents between the industry and the
public purse are catch royalties (i.e., taxes) on the one hand, and quota
auctions on the other. Most other publicly owned natural resources are
managed using one or the other of these methods (or both). There is
no reason why they could not be used in fisheries, although at present
most ITQ systems capture at most minimal rents for the government
(see Chapter 6).

The extent to which these methods tend to encourage resource-
conservation incentives depends on how the methods are applied. For
example, the auction of single-year catch quotas is unlikely to encourage
resource conservation, for obvious reasons, but auctioning longer term
quotas (e.g., 10 years) may do so. Some economists have recommended
that ITQs be permanent (to maximize the incentives for resource conservation), but in the absence of substantial catch royalties these would
constitute a give-away of the resource in perpetuity.

Another possibility is to use catch royalties alone, without any quotas. In theory this would lead to a taxed bionomic equilibrium, with all resource rents being taken as royalties. As far as I know, this idea has never actually been tried out. Weitzman (2002) argues that catch taxes are economically superior to quotas (including auctioned quotas) if the annual catch is uncertain. He assumes that the quotas must be specified in advance, so that individual fishing strategies may be suboptimal if the availability of fish is unexpectedly low, or high. The whole question of individual catch quotas needs more research; further discussion appears in Chapter 4. But this does not mean that the use of ITQs should be delayed until we know everything about them.

Community Quotas

One of the advantages of an ITQ-based management system is that it leads to decentralized decision making. Quota owners are motivated to cooperate with the management authority in setting regulations, and in ensuring compliance. Actual examples are discussed in Chapter 6.

Decentralization can also be achieved, at least in principle, through group quotas, allocated to various coastal communities, or other identifiable groups. These groups then decide how to allocate catches (or economic benefits) among their members. For example, community control of coastal resources has been used successfully in Japan for centuries (Yamamoto 1995). The system works well for sedentary species, such as shellfish, but has proven less successful for mobile species. Many examples of community management of common-property resources in general are described in Ostrom (1990).

1.7 Fishery Resources as Natural Capital

It is important to realize that the Gordon model of Fig 1.1 makes no explicit reference to the fish population itself. This figure might be mistaken for a direct input-output model—any given input of effort E apparently implies a given output (yield) of fish Y. This is not how effort and yield are actually related, in most fisheries. Instead, the catch rate from a population of fish depends both on effort and on the current size of the fish stock. A standard model (Schaefer 1954) for this relationship is

$$h = qEx \qquad (1.5)$$

where h is the harvest rate, E is effort, x is the current biomass of the fish population, and q is a constant called "catchability." We discuss this and other catch relationships in Chapter 2.

Catching fish will usually have an effect on the population, and this can be modeled by the differential equation

$$\frac{dx}{dt} = G(x) - h \tag{1.6}$$

where $G(x)$ represents the natural net growth rate of the population. Harvesting at a rate higher than this natural replenishment rate ($h > G(x)$) reduces the stock of fish over time. Conversely, harvesting with $h < G(x)$ permits the stock to increase over time. We also study these questions in detail in Chapter 2.

From an economic point of view, a population of fish can be considered as a form of capital—natural capital, in contrast to man-made capital. Both man-made and natural capital are capable of generating an annual production, or "yield," depending on the current size of the asset. But also (and this is highly significant) the size of the asset can be either increased or decreased, over time. Increasing the size of an asset is the same as investing in it; decreasing is the same as disinvesting. In determining the optimal size of an asset, one must take account of the potential yield and also of the costs and benefits of investment or disinvestment. The static Gordon model ignores the latter aspects of natural capital. As will be shown in Chapters 2 and 3, the implications of the Gordon model change dramatically when it is replaced by a dynamic model.

1.8 Multispecies Fisheries and Ecosystem Based Management

The practical determination of annual catch quotas (TACs) has usually been based on single-species models of population dynamics. The models have been criticized for ignoring important features of marine systems. It has been recommended that management decisions should be based instead on multispecies or ecosystem models (e.g., May et al. 1979; Mangel et al. 1996; Pikitch et al. 2004). Once this degree of model complexity is envisioned, it would seem appropriate also to include other real-world aspects such as spatial non-homogeneity and environmental fluctuations. Such models can only be realized in terms of computer simulations, which are certainly feasible with today's machines and software packages (Walters et al. 1997).

The next step would be to include similarly complex economic components in such ecosystem simulation models. This has been done for the case of open-access fisheries (Walters and Martell 2004), but apparently not yet for management systems involving ITQs and related methods.

In this book I mostly rely on single-species biological models, although multispecies aspects and implications of spatial non-homogeneity are occasionally considered. The basic bioeconomic principles of managed fisheries still seem sufficiently poorly understood that attempts to construct and study ecosystem-socio-bioeconomic models is probably a bit premature. Eventually, no doubt, this unification will be achieved. Some case studies discussed in Chapter 6 suggest that multispecies ITQs may be quite successful from the economic viewpoint, provided that the ITQ system is sufficiently flexible.

1.9 The Role of Uncertainty

How on earth can a management agency specify an annual TAC with any accuracy, given that in most cases the fish population is not directly observable, the marine ecosystem is poorly understood, and environmental fluctuations are largely unpredictable? Fishery managers deal with these uncertainties on a regular basis, sometimes with great success, but occasionally with disastrous results. Added to these scientific uncertainties is the fishing industry's built-in bias towards large quotas. Any admission of scientific uncertainty has usually prompted the response "if you can't prove that the catches we need will harm the stock, you have no excuse for keeping the quota low." Fishermen can always plead economic hardship as a justification for high quotas. Fishery scientists have not always been successful in convincing decision makers (often politicians) to adopt sufficiently strong catch restrictions to prevent overfishing, even to the point of collapse of the fishery. Because of the history of failed management, however, a new philosophy based on precautionary principles seems to be emerging.

Risk Management

How is precautionary management to be implemented? The first step is to identify the main sources of uncertainty that may affect a given fishery. Next one tries to quantify these uncertainties, and to devise management strategies that minimize the risk of overfishing (particularly the risk of collapse), while maintaining a productive and profitable fishery.

Chapter 5 discusses a new approach to risk management in fisheries, using the methods of decision analysis. The latter is a well-developed set of techniques that are increasingly being applied to the management of uncertainty and risk in environmental and other systems (Morgan and Henrion 1990). In outline, decision analysis (sometimes called risk analysis, or policy analysis) consists of the following steps:

1. Specify the problem
2. Obtain the necessary data. Identify uncertainties in the data.
3. Identify management options and objectives.
4. Construct a model or models. Identify process, parameter and structural uncertainties in the model.
5. Test the models.
6. Run model simulations for each management option.
7. Evaluate the management options, in terms of expected benefits *and risks*.
8. Document the study and obtain peer review.

Some of these steps are beginning to be used in fisheries, particularly in the area of stock assessment. For example, Bayesian methods can be used to quantify uncertainty about stock levels, and also about model hypotheses (e.g., McAllister and Kirchner 2002). But how these results should be applied to actual policy decisions has remained unclear.

As explained in Chapter 5, decision makers need to determine their degree of risk avoidance, as part of the decision process. For example, it might be decided to choose a management strategy that maximizes expected net revenues over a 25-year horizon, subject to the requirement that the estimated risk of collapse of the population be kept less than 5% over that period. Monte Carlo simulations of the model(s) can then be used to estimate both the expected revenues and the probability of collapse, for a range of management strategies. Will future fisheries management follow some such path?

Institutional Change

It may be doubted whether present-day institutions for managing marine fisheries are sufficiently flexible to adopt the changes needed for a sustainable, profitable fishing industry. In Chapter 7 I suggest that present management institutions need to be broadened to include an economics/decision-analysis branch to work closely with the scientific branch. In addition, it will be necessary to establish a public overseer, to ensure that the nation's fishery resources are properly managed for profit and equity. It seems unlikely that the political arm alone can be relied upon for this purpose, although ultimate power and responsibility must always lie with the politicians.

1.10 The Future

The future of marine fisheries may be bright, at least for those stocks located within the 200-mile zones of coastal states. A gradual transition to so-called rights-based management systems such as ITQ systems, is already under way, and should result in cooperation among fishers, and between fishers and managers. The state will still have a major role to play, in terms of assisting and enforcing the profitable use of these resources. The state should also benefit, both directly and indirectly, from the vastly improved economic performance of its fisheries sector. Rather than wasting large amounts of taxpayer money in a doomed attempt to support a collapsing industry, the government will itself profit from an economic success story. All this may be complicated and difficult, but surely there is hope that a new paradigm of cooperative resource management can be realized.

2

Dynamic Bioeconomic Models

The bioeconomic fishery model discussed in Chapter 1 is a *static* model. The yield curve of Fig. 1.1 depicts sustainable annual yield as a function of effort E. Switching from one effort level to another does not instantly switch from one point on this curve to another. For example, a reduction of effort in a fishery currently at bionomic equilibrium will usually result in an immediate decrease in yield, rather than an increase as suggested by the static model. The increased yield only shows up after the fish stock has had a chance to recover from past depletions. Similarly, any increase in effort will normally result in an immediate increase in yield, but this may later be tranformed into a decreased yield, as the stock becomes further depleted. These dynamic changes are extremely important economically, and a viable bioeconomic theory must include them. The dynamic theory is a bit more complicated, but vastly more meaningful and useful, than the static (i.e., equilibrium) theory.

Natural resource stocks can be considered a form of "natural capital" (Jansson et al. 1994). The question of optimal resource use is therefore a problem of optimal investment in, and consumption of natural capital (Clark and Munro 1975). Our dynamic bioeconomic fishery model will make this connection clear. But first we examine the underlying biological model. (A different biological model will be discussed in Sec. 2.7.)

2.1 A Dynamic Model

The crux of dynamic modeling is the use of a *state variable* to represent the current state of the system in question. Here we use

$$x(t) = \text{fish population biomass at time } t \tag{2.1}$$

To be specific, we assume that biomass $x(t)$ is measured in metric tonnes, and time t in years. The following *general production model* has been used in the management of tropical tuna and other fisheries (Schaefer 1954):

$$\frac{dx}{dt} = G(x) - h(t) \qquad (2.2)$$

where $G(x)$ represents the natural growth rate of the population, and $h(t)$ is the harvest rate. The simplest useful form for the growth function is the logistic:

$$G(x) = rx(1 - x/K) \qquad (2.3)$$

where r denotes the intrinsic growth rate (/yr) of the biomass, and K is the carrying capacity (tonnes).

In the absence of harvesting, Eqs. (2.2) and (2.3) combine to give

$$\frac{1}{x}\frac{dx}{dt} = r(1 - x/K) \qquad (2.4)$$

This equation says that the per-capita growth rate of the population equals r at low population levels, and that the growth rate declines linearly as x increases. The term "density-dependent growth rate" is often used to describe this situation, whether or not the relationship is linear. A more general model, with nonlinear density dependence, is provided by the equation

$$\frac{1}{x}\frac{dx}{dt} = r(1 - x/K)^{\alpha} \qquad (2.5)$$

where $\alpha > 0$ is a skew parameter. Here $\alpha > 1$ gives a growth function $G(x)$ that is skewed to the left, relative to the logistic model; $\alpha < 1$ implies a right skew (Fig. 2.1). Fox (1970) discusses the estimation and use of such skewed models in fisheries.

For the logistic model, and its skewed form, we have

$$\lim_{t \to \infty} x(t) = K \quad \text{(if } x(0) > 0\text{)} \qquad (2.6)$$

These models thus imply a stable natural equilibrium at $x = K$. They also imply that the population will recover from overharvesting if harvesting is terminated short of actual extermination. In actuality, whereas some heavily depleted fish populations have in fact recovered satisfactorily under a fishing moratorium (e.g., Norwegian herring; see Bjørndal et al. 2004), other populations have recovered slowly or not at all (Hutchings 2000). Models of population dynamics that incorporate a minimum

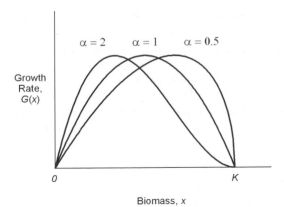

Figure 2.1 Growth function $G(x) = rx(1 - x/K)^{\alpha}$ for different values of α. (Graphs scaled to have the same maximum value.)

viable population size x_1 (with $dx/dt < 0$ for $x < x_1$) are said to involve critical depensation; see Clark(1990) and Liermann and Hilborn (2001).

Another class of population dynamics models uses discrete time steps $t = 0, 1, 2, \ldots$, which may be one year, or some other appropriate duration. The analogous equation to Eq. (2.2) is

$$x_{t+1} = F(x_t - h_t) \tag{2.7}$$

where x_t denotes the population size (biomass, or alternatively, number of adult fish) at the start of period t. As before, h_t is the annual harvest; the quantity $x_t - h_t$ is called the spawning escapement in period t, and x_{t+1} the resulting recruitment in the following period. Thus $F(\cdot)$ is the spawner-recruit function.

This discrete-time model is typically used for species with non-overlapping generations, such as the various Pacific salmon species. In this example, the traditional form of $F(\cdot)$ is due to Ricker (1954):

$$F(S) = Se^{r(1-S/K)} \tag{2.8}$$

Here r and K have the same interpretation as in the logistic model of Eq. (2.3). In particular $S = K$ is an equilibrium population size (with zero harvesting), since we have

$$F(K) = K$$

In contrast to the logistic model, this equilibrium is not necessarily sta-

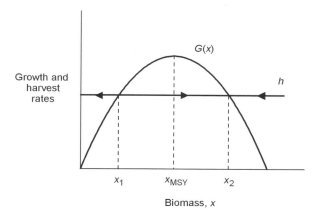

Figure 2.2 Constant-rate harvesting: x_2 is a stable equilibrium, and x_1 an unstable equilibrium.

ble, however. See Clark (1990, Ch. 7) for mathematical analysis of the Ricker model, and for bioeconomic models based on Ricker dynamics.

In this book we will work mainly with logistic-type models as in Eq. (2.2). These models provide a useful foundation for the economic analysis of commercial fisheries, which is our main objective.

Many other ways exist for elaborating dynamic models of marine populations. For example, models have been devised to incorporate age structure, spatial structure, genetics, multiple species, food webs, and oceanographic components. Large-scale stochasticity may also be included in the model. Walters and Martell (2004) brilliantly discuss many types of models used in fisheries biology, pointing out the dangers involved in overdoing model complexity.

The Effect of Harvesting

Let us begin with the case of constant-rate harvesting, $h = $ constant. Our dynamic model becomes

$$\frac{dx}{dt} = G(x) - h \tag{2.9}$$

Figure 2.2 shows how to deduce the consequences of this model. The $G(x)$ curve gives the rate of growth (or natural productivity) of the population, and the line at h gives the rate of harvest. Provided that the harvest rate h is sufficiently low, there are two equilibrium solutions to Eq. (2.9), x_1 and x_2, given by the solutions to the equation $G(x) = h$. The lower point x_1 is unstable, while x_2 is stable. (Note that $dx/dt < 0$

for $x > x_2$, whereas $dx/dt > 0$ for $x < x_2$ (if $x > x_1$).) If the population is initially at $x = K$, it is then fished down to x_2, where sustained fishing at rate h continues indefinitely.

It is enticing to ask what is the maximum harvest rate that can be sustained indefinitely? The answer, apparently, is given by

$$h_{\mathrm{MSY}} = \max G(x) \qquad (2.10)$$

and this implies an equilibrium biomass at $x = x_{\mathrm{MSY}}$. For the skewed logistic, $G(x) = rx(1 - x/K)^\alpha$, this gives

$$x_{\mathrm{MSY}} = \frac{K}{1 + \alpha} \qquad (2.11)$$

Thus $x_{\mathrm{MSY}} = K/2$ for the non-skewed logistic (Fig. 2.2), and $x_{\mathrm{MSY}} > K/2$ if the curve is right-skewed, etc. Maximum Sustained Yield (MSY) was long considered as the ideal objective of renewable resource management, including fisheries, presumably on the basis of the above (or similar) model (Larkin 1977).

It seems not always to have been noticed that x_{MSY} is an *unstable* equilibrium (Beddington and May 1977). The slightest deviation of h above h_{MSY} will push $x(t)$ below x_{MSY}, and eventually lead to extinction if maintained indefinitely. (Draw the h line just above the graph of $G(x)$ in Fig. 2.2, to see this.) In actuality, many marine fisheries have gone through a protracted period of overfishing, often resulting in extreme stock depletion, if not actual extinction. It would be oversimplistic to attribute most such instances of overfishing to errors in estimating MSY, although overly optimistic forecasts of sustainable yield have in fact occurred (e.g., Myers et al. 1997). Other aspects of the depletion and collapse of managed marine fisheries are discussed later in this chapter, and elsewhere in the book.

Reference Points

Surely, one would think, the catch rate h should be reduced when the stock falls to a low level, say $x < x_{\mathrm{MSY}}$. In other words, h should be some specified function of x:

$$h = h(x) \qquad (2.12)$$

The current terminology for this is *reference points* (Hilborn 2002), the idea being that $h(x)$ will typically have two thresholds, $x = a$ and b, as shown graphically in Fig. 2.3.

Any such harvest function $h(x)$ will in theory result in a stable equi-

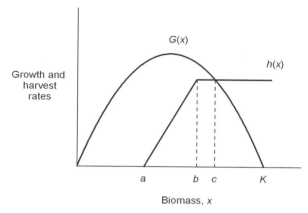

Figure 2.3 A reference-point harvesting strategy $h(x)$.

librium at $x = c$, and sustainable harvest rate $h = G(c)$. The reference-point strategy is an example of a *feedback control function*, meaning that the control variable (h) in Eq. (2.9) is a function of the state variable (x). (The reason for having a gradual reduction in $h(x)$ for $a < x < b$, rather than a cut-off at $x = b$, is that, in practice, the stock x may undergo random fluctuations. It would be highly undesirable to the fishermen if such variations caused the annual catch quotas to be turned completely off and on from one year to the next. Indeed, such on-off switching will also result merely from annual variations in estimated stock abundance, apart form any real changes. A reference-point strategy as shown in the figure reduces the impact of such variations.)

To implement a reference-point feedback strategy, it is clearly necessary to obtain regular estimates of the current stock abundance x. Such estimates may be expensive to obtain, and fairly inaccurate at best. Hilborn (2002) discusses various practical problems associated with reference points. Whether a fishery can be sustainably exploited over long periods without some such strategy based on continuing stock assessment remains problematic. We will return to this question later.

It is worth pointing out here that the management strategies discussed so far have zero economic content. They are single-mindedly concerned with achieving maximum yield from the resource, in a safe and sustainable manner. But the history of managed fisheries has been characterized by ongoing disagreement between fisheries scientists and the fishing industry. Almost without exception, the fishermen seem to think that the recommended catch quotas are unnecessarily small, especially when the fishermen are having no difficulty finding large concentrations of fish. The fact that maintaining these concentrations may be necessary for breeding purposes, and hence also necessary for the long-term sustainability of the fishery, simply does not seem to get through.

Is there in fact some rationale for this behavior? Do the fishermen have different incentives than what the scientists are assuming? Or is the fishing industry somehow organized ineffectively, in a way that forces fishermen to overfish? Much of the rest of the book is concerned with analyzing the economic incentives of fishermen, and how these incentives interact with resource management strategies.

But first we consider some additional aspects of our dynamic model of the fishery.

Constant-Effort Harvesting

We will now consider a different model assumption, namely that the harvest rate h is always proportional to effort E and to the population biomass x:

$$h = qEx \qquad (2.13)$$

Here effort is measured in units appropriate to the fishery in question, a common example being the number of standardized vessel units (SVU) actively fishing at a given time. Units of q, called the *catchability coefficient*, are thus (/SVU year). Intuitively, q specifies the fraction of the current biomass captured in one year by one SVU. Equation (2.13) is called the Schaefer catch equation (Schaefer 1954). We will discuss the implications of the Schaefer equation in some detail.

(A word here on the choice of units: I am using time units of years merely for convenience—and also because I am concerned with long-term consequences. This does not mean that E in Eq. (2.13) is assumed to remain constant for the entire year. For example, if 50 standard vessels fish for one day, with $x = 100,000$ tonnes and $q = .002/\text{SVU}$ yr, then the day's catch is $h\Delta t = qEx\Delta t = (.002) \times 50 \times 10^5 \times (1/365) = 27.5$ tonnes.

In fact, since our basic dynamic model is expressed as a differential equation, Eq. (2.14) below, h in Eq. (2.13) represents an instantaneous harvest rate, which can vary over the year even though time units are given as years. In practice, discrete-time models are normally used in fisheries, but these models involve rather messy formulas that, while familiar to fisheries scientists, tend to obscure the economic theory.)

Although the concept of fishing effort (E) is basic to fisheries science, providing a consistent definition of "effort" is by no means straightforward. The above statement, that effort is measured in terms of fishing activity (e.g., SVU), in fact defines what is usually called nominal effort. Nominal effort is thus the input to the fishing process.

In a second interpretation, effort is defined as the rate at which sea water is screened for fish (Rothschild 1972). Let us refer to this as physical effort. Assuming that E in Eq. (2.13) stands for nominal effort, this equation then incorporates two critical assumptions. First, physical effort is assumed proportional to nominal effort. Second, the density of fish throughout the fishing area is always directly proportional to total stock abundance x. Both assumptions are of questionable validity. We will discuss the second assumption in some detail later.

Assuming the unskewed form of the logistic growth model, $G(x) = rx(1 - x/K)$, our dynamic model, Eq. (2.2), now becomes

$$\frac{dx}{dt} = rx(1 - x/K) - qEx \tag{2.14}$$

As can be seen from Fig. 2.4, constant-effort harvesting, $E = $ constant, results in a stable equilibrium at $x = \bar{x}$, and a sustained harvest rate $\bar{h} = qE\bar{x}$, provided $qE < r$.

In our previously introduced terminology, the equation $h = qEx$ is a special case of feedback control, but now the control variable E is itself assumed to be a fixed constant, not a function of x. We can calculate the value of the equilibrium biomass $\bar{x}$ as a function of E, by setting $dx/dt = 0$ in Eq. (2.14):

$$rx(1 - x/K) - qEx = 0$$

which implies that either $x = 0$ or $x = \bar{x}$, where

$$\bar{x} = K(1 - \frac{qE}{r}) \quad \text{if } qE < r \tag{2.15}$$

We see that $\bar{x}$ is a stable equilibrium if $qE < r$, while $x = 0$ is unstable— see Fig. 2.4. If $qE \geq r$ then $x = 0$ becomes a stable equilibrium. Letting $\bar{x}$ now denote the stable equilibrium, we have in addition to Eq. (2.15).

$$\bar{x} = 0 \quad \text{if } qE \geq r \tag{2.16}$$

Note that qE is the fishing mortality rate, and r the intrinsic growth rate, of the population. As Eq. (2.16) shows, the population would be fished to extinction by constant-effort harvesting, if the fishing mortality rate exceeds the intrinsic growth rate. Otherwise (with $qE < r$)

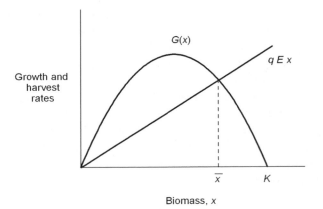

Figure 2.4 Constant-effort harvesting, $h = qEx$. The biomass level $\bar{x}$ is a stable equilibrium.

constant-effort harvesting results in a positive equilibrium biomass $\bar{x}$, with sustained yield $\bar{h} = qE\bar{x}$.

Next, from the equation $h = qxE$ we can now deduce the equilibrium yield-effort curve for our dynamic Schaefer model. From Eq. (2.15) we obtain

$$\bar{h}(E) = q\bar{x}E = \begin{cases} qKE(1 - qE/r) & \text{if } qE < r \\ 0 & \text{if } qE \geq r \end{cases} \tag{2.17}$$

This curve is the same as the yield-effort curve of the Gordon model (Fig. 1.1). Maximum sustained yield occurs at E_{MSY}, where

$$E_{\text{MSY}} = \frac{r}{2q} \tag{2.18}$$

(i.e., half the effort level corresponding to stock extinction). The value of h_{MSY} is obtained by combining Eqs. (2.11) with $\alpha = 1$, and (2.18):

$$h_{\text{MSY}} = qx_{\text{MSY}}E_{\text{MSY}} = rK/4 \tag{2.19}$$

The same calculation can be carried out for the skewed logistic model, the result being

$$\bar{h}(E) = qKE\left(1 - \left(\frac{qE}{r}\right)^{1/\alpha}\right) \quad (qE \leq r) \tag{2.20}$$

This is a skewed version of the (parabolically shaped) Gordon model of Fig. 1.1. Sustained yield is maximized at E_{MEY}, where

$$E_{\text{MEY}} = \frac{r}{q} \left(\frac{\alpha}{\alpha + 1} \right)^{\alpha} \tag{2.21}$$

For $\alpha > 1$ we have $E_{\text{MEY}} > r/2q$, and vice versa. Further analysis of the skewed model is left to the reader.

Thus we have demonstrated that our dynamic Schaefer model, with constant-effort harvesting, contains the static Gordon model as its set of equilibrium yield-effort pairs.

Predictions of the Constant-Effort Model

The constant-effort dynamic model (non-skewed case) has several appealing predictions from the management viewpoint. First, the model implies that MSY occurs when $qE = \frac{1}{2}r$, i.e., fishing mortality equals one-half of the population's intrinsic growth rate. This is an attractive rule of thumb.

Second, a constant-effort strategy results in a stable equilibrium, provided that fishing mortality is lower than the intrinsic growth rate. Overfishing to extinction does not occur unless effort exceeds E_{MSY} by at least 100%.

Third, natural stock fluctuations do not have a drastic effect on the catch rate, which remains proportional to x.

Fourth, any change in stock abundance is immediately signaled to managers, by the catch per unit effort (CPUE) statistic CPUE $= h/E$, because x is proportional to h/E. This implies that necessary alterations to the effort level E can be made when this signal so indicates.

Finally, the relationship between effort level E and sustainable yield $\bar{h}(E)$ is continuous, implying that small changes in effort will result in small changes in yield. This in turn implies that a gradual trial-and-error approach to management is appropriate. If effort E is inadvertently allowed to expand beyond E_{MSY}, this will eventually result in a minor decrease in annual catches, which can be corrected by reducing E slightly.

Clearly we need to know whether these appealing predictions are generally valid, or whether they are highly model specific. We now turn to this question.

2.2 Robustness of the Schaefer Model

The main theme of this book is the use of bioeconomic models to understand and predict the fishing industry's response to alternative management strategies. We therefore need to ask whether our predictions are strongly model dependent. How can one be sure that the model is OK? Are the predictions robust even if the model is seriously wrong? These are profound questions that may not have simple answers. Yet they must be asked—the alternative is ... well, the alternative is the sorry picture that we see today in global fisheries management. Or at least, modeling failure has arguably been an important part of the picture.

Depensation

The skewed-logistic curves shown in Fig. 2.1 all have the property that the per-capita growth rate $G(x)/x = r(1 - x/K)^\alpha$ is strictly decreasing as x increases. This is called *compensation*, the idea being that per-capita growth of the population decreases as the size of the population is increased, for example because of decreased availability of food, per individual. To state this differently, compensation means that per-capita growth rate increases as the population is reduced, for example by fishing. Some such process must occur, otherwise sustained harvests would not be possible.

However, for some marine species it has been shown that per-capita growth decreases as x is reduced below some critical level x_{dep}. This is called *depensation*. Two proposed causes of depensation are, first, reduced breeding success at low population density, and second, increased relative predation rates on small populations (see Liermann and Hilborn 2001).

An extreme type of depensation, called critical depensation, occurs if the growth rate $G(x)$ becomes negative for sufficiently small values of x. Whether critical depensation in fact occurs in any marine species seems to be unknown, but it is true that some severely depleted fish populations have shown little signs of recovery after fishing was curtailed (Hutchings 2000), an outcome that is consistent with the hypothesis of critical depensation.

Besides being a consequence of behavioral changes affecting single species, depensation can also arise from ecosystem processes. For example, Walters and Kitchell (2001) suggest that the phenomenon of "cultivation-depensation" may be common among large predatory

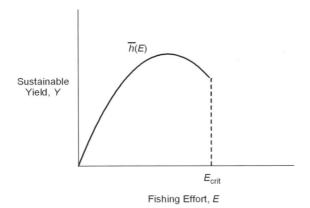

Figure 2.5 The sustained-yield-effort curve $Y = \bar{h}(E)$ under depensation.

species of fish. A reduction of the adult population of predators can allow their prey species to proliferate and compete with or prey upon juveniles of the targeted predator population. Thus depletion of adult predators may lead to a new system equilibrium characterized by permanently reduced abundance of the predator population.

The sustained-yield-effort curve $Y = \bar{h}(E)$ in the case of depensation (critical or not) is remarkably different from the case of compensation. An example is shown in Fig. 2.5. The curve has a sudden discontinuity at a certain critical effort level E_{crit}, which is to the right of E_{MSY}. Technically speaking, this discontinuity results from a cusp catastrophe in the dynamic model

$$\frac{dx}{dt} = G(x) - qEx$$

(To see how this discontinuity arises, redraw Fig. 2.4 for the case in which $G(x)$ is depensatory. How is E_{crit} determined from this graph?) We will not delve into the mathematical details here (see Clark 1990, pp. 18ff).

In practical terms, a fishery that exploits a population with depensatory dynamics may unexpectedly switch from sustainable harvesting to an extinction or near-extinction mode as effort is increased beyond the critical level. Furthermore, reducing effort to its previously sustained level may not return the fishery to its former sustained equilibrium. Instead, the stock may continue to decrease, possibly towards extinction. Only if effort is drastically reduced will there be any hope for rebuilding the stock. These predictions, which are clearly in sharp contrast to those of the standard Schaefer model of Fig. 2.4, are typical of catastrophic dynamics.

Precisely this mechanism—catastrophic decline resulting from depensatory population dynamics—has been suggested as an explanation for the collapse and non-recovery of Atlantic cod stocks off Newfoundland, by Myers et al. (1997). These authors hypothesize that decreased breeding success is the mechanism underlying depensation in the cod population. If correct, their analysis indicates that an almost complete moratorium may be required if cod stocks are to be rebuilt. In fact, although a declared moratorium on cod fishing is currently in effect, it appears that substantial illegal catches may still be occurring. Such catches could, at least in theory, prevent the stocks from ever recovering from their 1991 collapse.

For most, if not all marine populations the likelihood of depensatory dynamics (not to mention other complexities of ecosystem dynamics) is completely unknown. Thus the collapse and non-recovery of heavily fished populations, far from being surprising, should perhaps be considered as normal events.

The next question is, how then should fisheries be managed in view of the likelihood of unpleasant "predictable surprises"? We will defer the discussion of this question to Chapter 5, but we can at least state here that risk management in fisheries will probably require a much greater level of precaution than has been the custom until now. Models that have been estimated using data from a large and healthy fish population do not provide any information about the stock's probable dynamics after depletion has occurred. Keeping careful tabs on the population size, through regular stock assessment, is crucial.

An Alternative Catch Equation

Next we consider the role of the Schaefer catch equation

$$h = qEx \tag{2.22}$$

in terms of model robustness. We need to ask, what's behind this formula? Could it be of questionable validity?

The answer is that there are common and important situations where Eq. (2.22) is badly flawed. I will discuss two such situations pertaining to the role of x, and another to the role of E.

Note first that Eq. (2.22) assumes that as the fishery develops, the catch rate per unit effort h/E, will decrease in proportion to current stock abundance. For this to be true, it must be the case that the average density of the population over the fishing areas drops in direct proportion to total stock abundance. This simple observation suggests that Eq. (2.22) may be quite special. Here are two cases in which this equation is likely to be inappropriate.

1. Mobile pelagic species. Some species of fish respond to a reduction in population size by concentrating their numbers into smaller areas, thereby maintaining a high local density. This natural behavior may be in response to food availability or predation. The resulting concentrations of fish may be readily located by fishermen. In the limiting case where local density remains relatively constant, an appropriate catch-effort model would be

$$h = qE \qquad (\text{if } x > 0) \qquad (2.23)$$

where q is again a constant (units: tonnes/SVU yr). Not surprisingly, this modification completely changes all our model predictions, as we shall see.

2. Immobile demersal species. Imagine a large fish population spread fairly evenly over a large area. Over any brief time span, draggers remove virtually all the fish in a small sub-area, but other fish do not migrate into the vacated spaces. If the fleet moves progressively through the whole area (for example, moving steadily farther offshore), the catch per vessel day will remain almost constant as the stock declines. Again this implies Eq. (2.23).

Thus two apparently opposite behavioral patterns lead to the same catch model. A third case, highly migratory species, also leads to the same model, if the migration includes a stage where fish are concentrated in a relatively small area. The only situation in which the Schaefer model, Eq. (2.22), could be a good approximation to reality occurs when the fish population responds to reduction in numbers by redistributing itself uniformly over the whole fishing area. Perhaps the Schaefer model should be considered the exception rather than the rule in fisheries biology. In any event, CPUE should never be used as a primary index of stock abundance unless there is strong evidence that the catch equation $h = qEx$ is approximately valid for the stock under consideration. This question is discussed in more detail later.

The equation $h = qE$ is also an extreme case, in the sense that it assumes no dependence of the catch rate on the stock size x. Other, less extreme equations will be considered briefly later.

Replacing $h = qEx$ by $h = qE$ in our dynamic model, Eq. (2.14), gives

$$\frac{dx}{dt} = rx(1 - x/K) - qE \quad (x > 0) \tag{2.24}$$

In this model, constant-effort harvesting is the same as constant catch-rate harvesting. None of the supposed advantages of constant-effort harvesting apply under these circumstances. Instead we have the following predictions. First, constant-effort harvesting is unstable unless the catch rate qE is kept well below h_{MSY}. Second, natural fluctuations in stock abundance may suddenly switch the fishery to an extinction trajectory, unless a feedback control strategy is in use. Third, no CPUE signal occurs to indicate depletion of the stock. In other words, constant-effort harvesting can be very dangerous if Eq. (2.23) is a better fit to reality than Eq. (2.22).

Next we consider the role of effort E in the equation $h = qEx$. Fishing mortality F is defined by the equation

$$h = Fx \tag{2.25}$$

Hence the Schaefer model equation $h = qEx$ amounts to assuming that $F \propto E$. In practice, the problem is to specify a measure of effort in such a way that indeed $F = qE$, and then to calibrate the value of q. In general this is not possible, but one can hope for some sort of approximation. Basically, one needs to determine average fishing mortality of one standard fishing unit, such as one standard vessel, one trap, etc.

But even if this calibration has been accomplished, there remains a major problem: The fishing fleet changes over time, as vessels become more powerful and more efficient. Thus q must be recalibrated regularly. This requires accurate, frequent stock surveys. Statistical variation in the estimates of q may be large enough to mask any trend. My conclusion from this discussion is that constant-effort harvesting is often over-rated, perhaps as the result of failure to realize the importance of the assumptions behind the standard catch model, $h = qEx$. Simple models can be valuable aids to understanding nature, but only if the potential errors involved in the simplification are clearly recognized.

Unfortunately, the solution of the problem of oversimplistic models is not necessarily to turn to more complex models, thought to represent the "real" world more realistically. Indeed, the misuse of overly complicated, statistically unjustifiable models is if anything a more common error than the misuse of simple models. In modeling, life is never easy.

CPUE Profiles

Consider a fishery that gradually reduces the biomass $x(t)$ of a given population. At each time t we suppose that the fishing vessels encounter fish with mean density $\rho(t)$. Catch per unit effort is proportional to density, $h = q\rho(t)E$, where q is constant. Figure 2.6 shows three general trajectory types in x, ρ space.

Observe first that Type II (linear) corresponds to the Schaefer equation $h = qEx$, in which case we have $\rho(t) \propto x(t)$. Here CPUE remains directly proportional to stock abundance throughout the history of exploitation.

For a Type I CPUE profile, catch per unit effort declines more slowly than the stock level, implying that the CPUE index progressively overstates current stock abundance. The modified catch relation $h = qE$ considered earlier provides an extreme example of a Type I profile, with $\rho(t) = q$ remaining constant throughout the development of the fishery. In practice, a curved trajectory as shown in Fig. 2.5 is doubtlessly more realistic. As noted above, the Type I CPUE profile pertains to many fish populations, including pelagic schooling, demersal and migratory species. Type I species are highly vulnerable to overfishing, and may collapse with little warning.

Type III species exhibit reverse characteristics to Type I. CPUE declines rapidly at first, even though the degree of reduction of the population may be relatively small. For example, in an analysis of CPUE for North Sea demersal fisheries in the period following World War II, Gulland (1964) suggested that the rapid decline in CPUE resulted from the initial post-war exploitation of areas of high stock density (which were presumably remembered from pre-war experience), and did not accurately portray the true change in overall abundance. Any relatively sedentary fish stock with variable initial density would tend to produce a Type III CPUE profile, provided that the fishermen are able to locate and exploit high-density subareas.

Yellowfin and other species of tuna may provide another Type III example. Some tuna biologists believe that early declines in CPUE, which are quite typical for tuna fisheries, may not signal proportional stock declines, but result instead from the fact that the tuna quickly learn to avoid the fishing gear, or other causes. If this idea is correct, recent

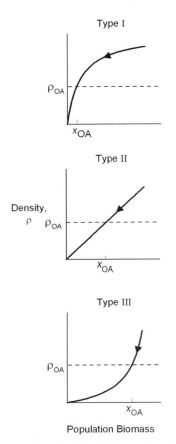

Figure 2.6 CPUE profiles $x(t)$, $\rho(t)$. The types are: (I) supralinear, (II) linear, (III) sublinear. The horizontal line represents the bionomic density ρ_{OA} under open-access fishing. The intersection of this line with the density trajectory determines the biomass equilibrium stock x_{OA}.

claims of severe depletion of tuna stocks, as reported by Myers and Worm (2003), and based on a linear CPUE profile model, may be overstated.

Table 2.1 summarizes these results.

The CPUE profiles in Fig. 2.6 imply a direct relation $\rho = \rho(x)$, and we have $h = q\rho(x)E$. This is sometimes expressed in terms of variable catchability $q(x)$, where $q(x) = q\rho(x)/x$ (q = constant), so that the catch equation becomes

$$h = q(x)xE \tag{2.26}$$

Table 2.1. *Implications of different CPUE profiles.*

Type	CPUE signal	Vulnerability to overfishing
I–supralinear	overestimates current stock level	High
II–linear	correctly estimates current stock level	Moderate
III–sublinear	underestimates current stock level	Low

Type II profiles have constant catchability, but the other types have variable catchability. In particular, a Type I CPUE profile implies increasing catchability as x is reduced (i.e., $q(x)$ is a decreasing function of x). The reverse holds for Type III profiles.

The management implications of nonlinear CPUE profiles are profound. The type of profile pertaining to a given fish population determines how CPUE is related to stock abundance. However, the type cannot be determined on the basis of CPUE data themselves. Fishery-generated CPUE data may therefore be suspect as an indicator of current stock abundance, unless it has been demonstrated that the fishery in question is approximately Type II (linear). Such demonstration requires an independent time-series of stock estimates for comparison with the CPUE time series. Such stock estimates can be obtained in various ways, including direct sampling of the population, tagging experiments, or the method called Virtual Population Analysis (VPA). The extensive literature on these methods is discussed by Walters and Martell (2004). These authors use the terms "hyperdepletion" and "hyperstability" to refer to Type I and Type III profiles, respectively.

What is the point of this lengthy (but incomplete) discussion of details of the Gordon–Schaefer dynamic model, you may ask. The main thrust of this book is bio-economic modeling, and our bio-economic models will often be based on the basic Gordon–Schaefer model, i.e., logistic general-production model plus linear CPUE model. This choice is convenient in terms of keeping the bio-economic theory fairly simple, but as always the dependence of model predictions on underlying model assumptions needs to be kept in mind. Although we cannot, for lack of space, perform a sensitivity analysis of every model relative to every assumption, we will occasionally examine the implications of alternative assumptions.

The above discussion of CPUE profiles is based on the idea that fish populations may not be uniformly distributed over the fishing area. Fishermen's tendency to favor the best fishing areas can lead to a nonlinear CPUE profile, and hence a nonlinear relationship between CPUE and abundance.

A nonlinear CPUE relationship can also result from random patchiness of the population. Patchiness implies that catchability q varies randomly over time as the fishermen encounter different densities of fish. Cooke and Beddington (1984) showed theoretically that a power law

$$h = cx^p E \quad (0 < p < 1) \tag{2.27}$$

approximately describes the catch-effort relationship under a variety of fishing techniques, including searching for schools, bottom trawling, and long-line fishing. The exponent p, called the catchability exponent, is a decreasing function of the variance in catchability q. Strongly patchy populations will thus exhibit strongly nonlinear Type I CPUE profiles. Thus Type I profiles may be common. As Cooke and Beddington say, "... a linear relationship between CPUE and abundance has rarely been demonstrated."

2.3 Dynamic Bioeconomics

To introduce some economics into our dynamic models, we now let p denote the price of fish (\$/tonne), and c the cost of effort (\$/SVU year). Total fleet net operating revenue R is then given (in \$/yr) by

$$R = ph - cE = (pqx - c)E \tag{2.28}$$

(We are now using the Schaefer catch model, Eq. (2.22); the alternative catch model of Eq. (2.23) will be discussed later.) In the dynamic model, both x and E are functions of time t. If $E = E(t)$ is specified, our dynamic model is

$$\frac{dx}{dt} = G(x) - qEx \tag{2.29}$$

$$G(x) = rx(1 - x/K) \tag{2.30}$$

$$R = (pqx - c)E \tag{2.31}$$

In the language of control theory, $x = x(t)$ is called the state variable, and $E = E(t)$ the control variable. Here we wish to consider two contrasting situations, the unregulated open-access fishery on the one hand,

and a dynamic, economically optimal control strategy on the other hand. As we shall see, for both situations the predictions of the dynamic model are quite different from those of the static model. What actually transpires in real-world fisheries is much better explained by the dynamic model. For example, predicting the behavioral response of fishermen to various regulations requires a dynamic modeling approach. Reliance on static bioeconomic models may result in inappropriate and unsuccessful policy recommendations.

The Unregulated Open-Access Fishery

We begin by modeling the dynamics of an unregulated open-access fishery. We will assume a fixed maximum effort level E_{max}:

$$0 \leq E(t) \leq E_{max} \tag{2.32}$$

Later, in Chapter 3, we will show how to predict the value of E_{max} from economic considerations.

If net operating revenue per vessel (R/E) is positive, i.e., if

$$pqx - c > 0$$

then fishing is profitable, and we expect that $E = E_{max}$. Likewise if this quantity is negative, we expect that $E = 0$. We conclude that

$$E_{OA}(t) = \begin{cases} E_{max} & \text{if } pqx(t) > c \\ 0 & \text{if } pqx(t) < c \end{cases} \tag{2.33}$$

where the subscript OA refers to Open Access. In other words, writing

$$x_{OA} = \frac{c}{pq} \tag{2.34}$$

we have

$$E_{OA}(t) = \begin{cases} E_{max} & \text{if } x(t) > x_{OA} \\ 0 & \text{if } x(t) < x_{OA} \end{cases} \tag{2.35}$$

Thus the fish population will be steadily reduced if $x(t) > x_{OA}$, or left to grow unfished if $x < x_{OA}$. The stock level $x = x_{OA}$ is therefore an equilibrium, termed the *bionomic equilibrium* stock level under open-access fishing. The level of effort $\bar{E}_{OA}$ that keeps $x(t)$ at x_{OA} is calculated by setting $dx/dt = 0$ in the dynamic model equation. This gives

$$rx_{OA}(1 - x_{OA}/K) = qx_{OA}\bar{E}_{OA} \text{ and } x_{OA} = c/pq$$

or

$$\bar{E}_{\text{OA}} = \frac{r}{q}(1 - \frac{c}{pqK}) \qquad (2.36)$$

(This result can also be obtained from the equilibrium yield expression, Eq. (2.17), i.e., $h = qKE(1 - qE/r)$. Combining this with $ph = cE$ immediately implies Eq. (2.36) again. This checks that our current assumption about E_{OA} in Eq. (2.35) coincides with the notion of bionomic equilibrium introduced in Chapter 1, Eq. (1.3).)

To complete this calculation, we need to consider the possibility that $c/pqK > 1$, which would seem to imply $\bar{E}_{\text{OA}} < 0$ in Eq. (2.36). However, since $x_{\text{OA}} = c/pq$, this would imply that $x_{\text{OA}} > K$, i.e., the open-access equilibrium is larger than the carrying capacity. In this situation the fishery is not economically viable at any feasible stock level, so in fact $\bar{E}_{\text{OA}} = 0$. Altogether we have

$$\bar{E}_{\text{OA}} = \begin{cases} \frac{r}{q}(1 - \frac{c}{pqK}) & \text{if } c < pqK \\ 0 & \text{if } c \geq pqK \end{cases} \qquad (2.37)$$

It is possible that $E_{\text{max}} < \bar{E}_{\text{OA}}$, i.e., that effort capacity E_{max} is too small to capture the sustained yield at $x = x_{\text{OA}}$. If so, the fishery would fail to reach bionomic equilibrium, and net revenues R would remain positive indefinitely. Although we will ignore this possibility here, we will allow for it in Chapter 3.

To review the above argument, our dynamic model of the unregulated open-access fishery assumes that fishing will occur at a certain rate ($E = E_{\text{max}}$) whenever fish are sufficiently abundant to provide catch rates that repay the costs of fishing. The harvesting of fish then reduces the fish population, at a rate given by Eq. (2.29), eventually reaching a stock level $x = x_{\text{OA}}$ below which the potential revenues from fishing would become negative. Hence fishing below x_{OA} is suppressed.

Once $x(t)$ has reached x_{OA}, a rather tenuous equilibrium is achieved. The natural growth rate of the population tends to cause $x(t)$ to increase, but this attracts fishing activities, which then again force $x(t)$ back towards x_{OA}. In this situation, fishermen continuously catch the sustainable yield $G(x_{\text{OA}})$, using a sustained effort level $E = \bar{E}_{\text{OA}}$.

This scenario is obviously a considerable abstraction. For example, any actual fish population is likely to undergo seasonal dynamics rather than continual growth $G(x)$. The fishery will then also take place on a seasonal basis. But it is still reasonable to suppose that fishing will persist as long as it remains profitable. Economic parameters, the price of fish, and cost of fishing, will determine an annual escapement biomass level x_{OA} as above. Fishing for the season will terminate when the stock has been fished down to x_{OA}.

A dynamic model can readily be constructed to fit this seasonal description—indeed, most models that are used in fisheries management do account explicitly for seasonal effects. Nevertheless we will stick with the above continuous-growth model, in order to retain analytic simplicity. (See Clark 1990, Ch. 7 for details of discrete-time, seasonal models.)

Dissipation of Economic Rent

Returning to our basic model, Eqs. (2.28)–(2.36), note that at bionomic equilibrium $x = x_{OA}$ we have

$$R = (pqx_{OA} - c)E_{OA} = 0 \qquad (2.38)$$

As in the static Gordon model (Fig. 1.2), net revenues from fishing are zero at bionomic equilibrium. The fishery-economics literature refers to this result as the "dissipation of economic rent" in the open-access fishery (e.g. Christy and Scott 1965). "Rent" is a technical term in economics, defined as the payment for the use of a resource. In the present setting we equate resource rent to the net revenue R in Eq. (2.31), i.e.,

$$\text{Rent} = \text{Net revenue} = \text{Revenue} - \text{Cost}$$

Rent is a flow, measured for example in \$/yr. Although rent is zero, or nearly zero, under open access conditions, an economically viable fishery is capable in principle of producing a positive sustained rent, which would constitute a positive long-term contribution to the economic welfare of society. The dissipation of rent in the open-access fishery therefore amounts to a dead loss of social welfare—a waste of valuable resources. Presumably this waste could be prevented somehow. Exactly how is the main theme of this book.

Foreshadowing our future discussion, let me say that preventing the dissipation of economic rent in commercial fisheries has proved to be a good deal more difficult than anticipated (Caddy and Seijo 2005). In fact, it was only during the writing of this book that I, with my colleagues, came to understand the full difficulties of preserving economic rents in multi-user fisheries. Exactly why this is true will only emerge slowly from the discussion.

It seems clear that many existing management programs have failed to produce economic benefits (rents) that are anywhere near the maximum that could be achieved. Indeed many management programs have probably resulted in substantial negative overall rents, if the costs of management are included in the accounts.

The Possibility of Extinction

The bionomic equilibrium stock level x_{OA} of Eq. (2.34) is proportional to the cost-price ratio c/p. Higher prices for fish, or lower costs of fishing, imply a greater degree of resource depletion under open access. But extinction of the fish population does not occur, no matter how low the cost–price ratio is.

What is the reason for this encouraging prediction? We can express fishing costs in a different way, by considering the cost of a unit harvest from a population of size x. We have

$$h\Delta t = qxE\Delta t = 1$$

(where Δt is the time required to catch one unit of fish). The resulting cost is $cE\Delta t = c/qx$. We write this as

$$\text{unit cost of fishing} = c(x) = \frac{c}{qx} \qquad (2.39)$$

Thus, for the Schaefer model, the unit cost of fishing becomes indefinitely large as the population tends towards extinction. This reflects the tacit assumption (discussed earlier) that the population always distributes itself evenly over the fishing grounds. Fish species that behave in this way are safe from biological extinction in a commercial fishery—provided that certain other tacit assumptions are also valid.

In particular, the Schaefer-logistic model assumes that the population in question does not have critically depensatory dynamics, for if this is the case, extinction would occur if $x_{OA} < x_{crit}$. Other biological mechanisms that may lead to the extinction, or near extinction, of heavily fished populations include the reduction in fecundity among young female fish (Berkeley et al. (2004); see also Myers and Mertz (1998)), and inter-species reactions (Walters and Kitchell 2001).

Alternative Models

How do the above formulas change for alternative forms of the basic model? If we assume a general growth function $G(x)$ and general catch-rate equation $h = q\rho(x)E$, our model becomes

$$\frac{dx}{dt} = G(x) - q\rho(x)E \qquad (2.40)$$

$$R(E) = (pq\rho(x) - c)E \qquad (2.41)$$

Bionomic equilibrium x_{OA} is now determined by the equation

$$\rho(x_{OA}) = \frac{c}{pq} \tag{2.42}$$

For the curves shown in Fig. 2.6 (or given by Eq. 2.26) we again obtain $x_{OA} > 0$. Extinction does not occur under open access unless $G(x)$ involves critical depensation (except for the limiting case of density independence, $h = qE$; in this case open-access fishing will exterminate the population if $pq > c$).

How serious is the threat of irreversible damage to a fish stock, under unregulated fishing? It seems unlikely that fishermen would deliberately search out the last fish of a given population, but this is not the only path to long-lasting commercial extinction. As already noted, complex ecosystem or behavioral processes may inhibit the recovery of severely depleted fish populations (Hutchings 2000; Walters and Kitchell 2001). Recovery may also be inhibited as a result of by-catches in a fishery directed at other species. Examples from the North Atlantic are described by Pauly and MacLean (2003).

Nonlinear Effort Costs

We next examine the concept of fishing effort in somewhat greater detail, concentrating on the decisions of individual vessel owners. Recall that the basic physical definition of fishing effort is the rate at which fishing gear screens the fishing area. In our previous models we assumed that effort could also be identified as the number of standard vessels operating at any given time. This measure, called nominal effort, allowed us to suppose that the cost of effort is directly proportional to (nominal) effort.

We now reformulate the effort model as follows. Let E_i denote the daily effort of vessel i. Daily effort can be varied in various ways, for example by fishing more hours per day, taking on more crew, running engines at high speed, and so on. Total fleet daily effort is given by

$$E = \sum_{i=1}^{N} E_i \tag{2.43}$$

where N is the number of participating vessels. We assume here that N is fixed; this convenient but artificial assumption will be relaxed in Chapter 3.

Effort costs may rise sharply when fishing activities are intense. We will now assume that

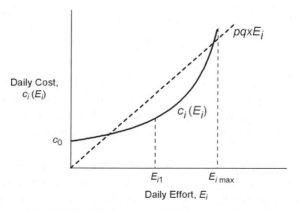

Figure 2.7 Daily effort costs $c_i(E_i)$ for an individual vessel.

$$\text{daily effort cost for vessel } i = c_i(E_i) \qquad (2.44)$$

with positive, non-decreasing marginal cost:

$$c_i'(E_i) > 0, \quad c_i''(E_i) \geq 0 \qquad (2.45)$$

(see Fig. 2.7). A special case, which was tacitly assumed in the basic model, is that

$$c_i(E_i) = c_i E_i \text{ for } 0 \leq E_i \leq E_{i\,\text{max}}$$

In this case we can think of marginal costs as jumping to $+\infty$ at $E_i = E_{i\,\text{max}}$.

Net revenue flow (\$/day) for the ith vessel is

$$R_i = pqxE_i - c_i(E_i) \qquad (2.46)$$

(where q now denotes catchability per day). To specify the individual optimal effort E_i we first define the "shut down" effort level E_{i1} by

$$c_i'(E_{i1}) = c_i(E_{i1})/E_{i1} \qquad (2.47)$$

(see Fig. 2.7). Then the optimal E_i is given by

$$\left.\begin{array}{ll} E_i = 0 & \text{if } pqx < c_i'(E_{i1}) \\ c_i'(E_i) = pqx & \text{if } c_i'(E_{i1}) \leq pqx \leq c_i'(E_{i\,\text{max}}) \\ E_i = E_{i\,\text{max}} & \text{if } c_i'(E_{i\,\text{max}}) < pqx \end{array}\right\} \qquad (2.48)$$

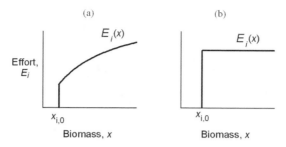

Figure 2.8 Individual effort response functions: (a) nonlinear effort cost; (b) linear effort cost.

(This can be seen by varying the slope of the line $R = pqxE_i$ in Figure 2.7.)

Equation (2.48) determines the ith vessel's response function $E_i = E_i(x)$, an increasing function of x, as illustrated in Fig. 2.8.

We have

$$E_i(x) = 0 \quad \text{if } x < x_{i,0}$$

where

$$x_{i,0} = c_i'(E_{i1})/pq \tag{2.49}$$

(Figure 2.8). Note that this agrees with the previous model, $x_{OA} = c/pq$, in the case that all N vessels are identical and have linear costs.

As in Eq. (2.43), the industry response function $E(x)$ is the sum of individual response functions

$$E(x) = \sum_{i=1}^{N} E_i(x) \tag{2.50}$$

Fishery dynamics are now

$$\frac{dx}{dt} = G(x) - qxE(x) \tag{2.51}$$

What about bionomic equilibrium? Recall that in the basic model, all vessels were assumed to be identical. A common shut-down level for each vessel, $x_{OA} = c/pq$, was therefore the model's predicted bionomic equilibrium. In the present model, vessel owners adjust their effort levels according to $E_i = E_i(x)$. Total fleet effort $E(x)$ in Eq. (2.50) is an increasing function of the stock level x—see Fig. 2.9. Bionomic equilib-

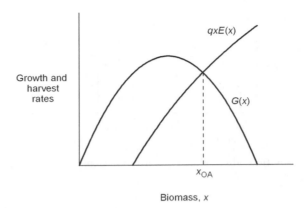

Figure 2.9 Bionomic equilibrium in the case of nonlinear costs.

rium x_{OA} is now determined by setting $dx/dt = 0$ in Eq. (2.51), and this implies that

$$G(x_{OA}) = qx_{OA}E(x_{OA}) \qquad (2.52)$$

Since the right side of this equation is increasing in x_{OA}, there will normally be a unique solution x_{OA} to the equation, as indicated in Fig. 2.9.

The predictions of this disaggregated, nonlinear effort model differ in two respects from the previous model. First, sustained net revenue $R = \sum_i R_i$ remains positive at bionomic equilibrium. However, this sustained revenue is likely to be much smaller than the overall optimum. Second, while some vessels continue to fish at bionomic equilibrium, other, less efficient vessels may be driven out of the fishery entirely. The latter vessels are those for which $x_{i,0} > x_{OA}$.

The disaggregated, nonlinear model, and its predictions, seem more in accord with real fisheries than the basic model. Every fishery has its "highliners"—fishermen who continue to make good money while others struggle to break even. Presumably the highliners are simply better, more efficient fishermen. But the more such fishermen there are, the greater will be the degree of stock depletion under open-access conditions.

We will not persist with this disaggregated, nonlinear model for the rest of this chapter and the next (although an aggregated nonlinear model will be used in Sec. 2.6). The disaggregated model will however become essential for the analysis of management strategies in Chapter 4.

2.4 Dynamically Optimal Fishing

Our next problem is to define what we mean by optimal fishing in the dynamic, or time-dependent sense. What time schedule of fishing effort $E(t)$—and hence of catch rates $h(t)$—is best, in some sense? I will initially take the traditional economic approach to this question. Imagine that the entire fish resource is owned by a single firm (or by the government), which possesses complete knowledge of and control over the resource. Economic theory tells us that the owner will manage the fishery so as to maximize the total *discounted present value* of resource rents $R(t)$:

$$\underset{\{E(t)\}}{\text{maximize}} \int_0^\infty e^{-\delta t} R(t)\, dt \qquad (2.53)$$

where δ denotes the discount rate. Here $R(t)$ is given by Eq. (2.31), and $E(t)$ satisfies Eq. (2.32), and the biomass $x(t)$ satisfies Eq. (2.29), exactly as in the open-access model.

In Clark (1990) I give a fairly comprehensive explanation of Eq. (2.53). Here just let me point out that the integral in Eq. (2.53) adds up rents $R(t)$ over all future time $t \geq 0$, discounting rents received at time t by the discount factor $e^{-\delta t}$. Discounting future revenues is the normal procedure for financial transactions of all kinds. The discount rate is the same as an interest rate. Discounting implies that the decision maker (investor) prefers current revenue to future revenue to some extent, determined by other investment opportunities. Discounting is an extremely important consideration in the field of resource and environmental conservation (Hotelling 1931; Ciriacy-Wantrup 1972; Clark 1990; Sumaila and Walters 2005).

Our mathematical problem is now to determine the effort schedule $E(t)$ (or catch schedule $h(t)$) that maximizes the present value integral in Eq. (2.53). Intuitively, given that future revenues are discounted, it may seem that the resource owner would prefer to fish heavily in the present and accept smaller catches in the future. This is correct, but only to a degree. The degree is determined by the tradeoff between present and future revenues. This tradeoff, in turn, is influenced by all the economic and biological parameters of our model.

Mathematically speaking, the maximization problem in Eq. (2.53) is problem in Optimal Control Theory—a difficult branch of mathematics. The present example, however, can be solved by fairly elementary means. Of course this elementary solvability comes at a cost, namely the large number of simplifying assumptions used in the model. Later we will briefly describe the consequences of altering some of these assumptions.

Box 2.1 gives an outline of the derivation. Here I will describe and discuss the optimal solution.

Box 2.1 The dynamically optimal harvest strategy

Our dynamic optimal control problem is:

$$\underset{\{h\}}{\text{maximize}} \int_0^\infty (p - c(x))h(t)e^{-\delta t}dt \qquad (2.54)$$

where

$$h(t) = qE(t)x(t) \qquad (2.55)$$

and

$$c(x) = \frac{c}{qx} \qquad (2.56)$$

subject to

$$\frac{dx}{dt} = G(x) - h(t), \quad x(0) = x_0 \qquad (2.57)$$

$$0 \le h(t) \le h_{\max} = qx(t)E_{\max} \qquad (2.58)$$

We define

$$z(x) = \int_{x_{\text{OA}}}^x (p - c(x))\,dx \qquad (2.59)$$

so that

$$\frac{dz}{dt} = (p - c(x))\frac{dx}{dt} \qquad (2.60)$$

Substituting from Eq. (2.57) into (2.54), and using (2.60), we obtain

$$J = \int_0^\infty (p - c(x))G(x)e^{-\delta t}\,dt - \int_0^\infty \frac{dz}{dt}e^{-\delta t}\,dt$$

Box 2.1. *continued*

where J denotes the integral in Eq. (2.54). By integration by parts, we have

$$\int_0^\infty \frac{dz}{dt} e^{-\delta t} \, dt = ze^{-\delta t}|_0^\infty + \delta \int_0^\infty ze^{-\delta t} \, dt$$

$$= \delta \int_0^\infty ze^{-\delta t} \, dt - z(0)$$

and this implies that

$$J = \int_0^\infty [(p - c(x))G(x) - \delta z(x)]e^{-\delta t} \, dt + z(0) \qquad (2.61)$$

The optimal harvest (and effort) strategy is now almost apparent. The stock biomass x should be adjusted, through harvesting, so as to maximize the expression in square brackets in Eq. (2.61). By differentiation, this implies that the optimal target equilibrium biomass $x = x_\delta$ satisfies Eq. (2.63) in the text. Moreover, because of the discount factor $e^{-\delta t}$, this adjustment should be made as rapidly as possible. This implies that

$$E(t) = \begin{cases} E_{\max} & \text{if } x > x_\delta \\ G(x_\delta)/qx_\delta & \text{if } x = x_\delta \\ 0 & \text{if } x < x_\delta \end{cases} \qquad (2.62)$$

The Optimal Equilibrium Biomass

The optimal dynamic fishing strategy for the present model, that is, the effort strategy $E(t)$ that maximizes the discounted present value in Eq. (2.53), turns out to be the fishing strategy that adjusts the initial biomass $x(t)$ as rapidly as possible towards a certain long-term target equilibrium biomass $x = x_\delta$. Specifically, x_δ is the solution of the following equation:

$$G'(x_\delta) - \frac{c'(x_\delta)G(x_\delta)}{p - c(x_\delta)} = \delta \qquad (2.63)$$

Table 2.2. *Target blue whale population levels, computed from Eq. (2.64).*

Discount rate δ (/yr)	Optimal target population x_δ (whales)
0	76,000
1%	61,488
3%	33,673
5%	12,758
10%	3,853
20%	2,647
∞	2,000

where

$$c(x) = \frac{c}{qx}$$

Typically, Eq. (2.63) has a unique solution x_δ (examples below), which is a decreasing function of the discount rate δ.

Let me use the Antarctic blue whale fishery as an illustration (Clark and Lamberson 1982). I assume a logistic growth function $G(x)$, with $K = 150,000$ blue whales and $r = .05$/yr. In this case Eq. (2.63) can be solved explicitly for x_δ:

$$x_\delta = \frac{K}{4} \left[\frac{x_{OA}}{K} + 1 - \frac{\delta}{r} + \sqrt{\left(\frac{x_{OA}}{K} + 1 - \frac{\delta}{r} \right)^2 + \frac{8 x_{OA} \delta}{Kr}} \right]$$

(2.64)

Here, as before, we have $x_{OA} = c/pq$.

Given that blue whales were hunted down to about 2,000 whales by the early 1960s, let me assume that in fact $x_{OA} = 2,000$.

Table 2.2 gives the numerical values of x_δ. In this example, the economically optimal harvest strategy is extremely sensitive to the discount rate, differing little from open-access bionomic equilibrium at values of δ as low as 10% per annum. The reason for this will be explained later.

In the present model, it is never optimal to completely exterminate blue whales, because whaling becomes unprofitable for $x < 2,000$ whales.

In actuality, financially "optimal" extinction of Antarctic blue whales is a distinct possibility for at least two reasons. First, if whaling continues for other species in the Antarctic (fin, sei, humpback and minke whales have all been hunted), the few remaining blue whales may occasionally be encountered. Being the largest and most valuable of the baleen whales,

Table 2.3. *Optimal target blue whale population x_δ as a function of the discount rate δ (case of zero, or stock-independent costs).*

Discount rate δ (/yr)	Optimal target population x_δ (whales)
0	75,000
1%	60,000
2%	45,000
3%	30,000
4%	15,000
$\geq 5\%$	0

blues would probably be harpooned. A blue-whale model with zero cost of effort might be appropriate in this situation. Table 2.3 shows the optimal equilibrium population levels x_δ calculated from Eq. (2.64) with $c = 0$.

If you haven't seen this before, I assume you are shocked. Hunting of blue whales to the point of extinction may seem likely under open-access, but could this really be economically optimal in some sense? What about other species? Are whales special in some way?

As a more recent example of discounting, we cite the case of a major fishery for snapper in New Zealand (Hilborn et al. 2003). A proposal to rebuild the snapper stock was challenged (successfully?) in the courts. The proposal called for a 40% reduction in catches over a 20 year period, which would then allow for an 8% increase in catches, relative to current levels. Proposals of this nature are almost always resisted by the fishing industry. Unless stocks are so depleted that there is a high risk of collapse, it seems hard to justify such recovery projects. In the snapper example, the rate of interest earned on the recovery "investment" would have been 0.8% per annum (neglecting the cost savings from having a larger fish stock).

Other questions raised by these examples include:

1. Are discount rates unusually high in fisheries? Why?
2. Do fishermen explicitly consider the discount rate in making decisions?
3. What is a socially optimal discount rate?
4. What are the effects of other economic parameters?
5. How can a fishery be managed in a socially optimal manner?

What is "optimal"?

First, we need to stress that the word "optimal" is being used here in a technical sense. The optimal harvest strategy, by definition, is the strategy that maximizes the present-value function, Eq. (2.53), subject to the model equations specified earlier. Thus the term "optimal" does not carry any moralistic implications (a point that is sometimes misunderstood).

Firms (and individuals) do discount the future, and bioeconomic models must take account of this fact. Furthermore, maximizing the present value of returns on investments, including investments in natural capital, is widely accepted as a socially optimal strategy—provided that the discount rate is itself socially optimal.

Discounting and Conservation

Exactly why is it that discounting the future is antithetical to resource conservation? (Ciriacy-Wantrup (1972) called this the fundamental law of conservation economics.) The intuitive answer is that conservation implies a concern for future benefits, whereas discounting places greater emphasis on current consumption, depending on the magnitude of the discount rate. High discount rates by resource users imply that they strongly prefer immediate short-term gains to delayed long-term returns. What effect this preference has in the case of a biological resource depends on the biological growth rate of that resource. (Note that x_δ in Eq. 2.64 depends on the ratio δ/r of the discount rate δ to the intrinsic growth rate r.) One reason that the whale model predicts severe economically "optimal" depletion of blue whale stocks at moderate interest rates (5–10% per annum) is that the blue whale population itself has a low growth rate (maximum of 5% per annum). In simple language, money-in-the-bank at interest rate δ grows faster than whales-at-sea, if $\delta > r$. The only thing that prevents extinction of blue whales is that the cost of catching a blue whale becomes greater than the value of the whale, if $x < 2,000$ whales. As Table 2.3 shows, if this cost condition were not true, extinction would be economically optimal for $\delta > r$. Surely the fact that Japan still argues today for the right to continue catching blue whales (ostensibly for research purposes) must have something to do with discounting the future! The current market value for the meat of one blue whale is probably in excess of US $100,000.

So, in this sense whales are exceptional. Few other marine species have such low intrinsic growth rates, although other K-selected species such as sea turtles and sharks may. Not surprisingly, many sea turtle and shark populations are currently severely depleted, most species being now classified as endangered.

The implications of discounting are not limited to species with low intrinsic growth rates, however. For example, many economically important species, such as cod, halibut and other groundfish, have low individual post-larval growth rates. These species can attain large body sizes under natural conditions, but under commercial harvesting the average size of fish in the population is often greatly reduced. Indeed, with unregulated fishing, the average size of fish in the catch may be a very small value; this is sometimes called growth overfishing. But some degree of long-term size reduction would also be consistent with an economic optimization model, particularly at a high discount rate (see Sec. 2.7).

But do fishermen and fishing companies operate under high discount rates, and why? Paramount among the reasons for high discount rates in open-access resources is the lack of assurance that the resource stock will persist at a profitable level into the future at all. Indeed, as we know, under unregulated open access, a fish stock will be reduced to bionomic equilibrium, at $x = x_{OA}$. In this situation each fisherman is forced in essence to entirely discount the future, so that $\delta = +\infty$. Referring to Eq. (2.63), we see that as $\delta \to +\infty$ we have $p - c(x_\delta) \to 0$ (this being the only way that the left side of Eq. (2.63) can become infinite), and therefore

$$\lim_{\delta \to +\infty} x_\delta = x_{OA} \qquad (2.65)$$

because $p - c(x_{OA}) = 0$ by definition. Note that this holds numerically for the blue whale example in Table 2.2. We thus conclude that the unregulated open-access fishery is equivalent to the economically optimal fishery with infinite discounting, an observation first provided by Scott (1955). We will return to this point in a moment.

Zero Discounting

For completeness, we consider the case of zero discounting, $\delta = 0$. In this case, future revenues are weighted equally with current revenues. (Since the integral in Eq. 2.53 diverges for $\delta = 0$, we have to treat this case as the limit for $\delta \to 0$.) We expect that the optimal harvest strategy in this case will maximize sustainable net revenues. For sustainable yield we have $h = G(x)$, so that

$$R = (p - c(x))h = (p - c(x))G(x)$$

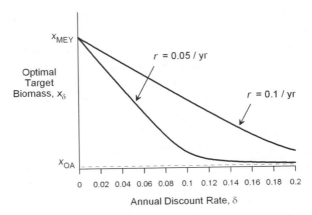

Figure 2.10 The optimal target biomass level x_δ as a function of the discount rate δ, for two values of the intrinsic growth rate r. (Data from the blue whale example.)

To maximize this expression we must have

$$(p - c(x))G'(x) - c'(x)G(x) = 0 \qquad (2.66)$$

and this is the same as Eq. (2.63) with $\delta = 0$. We therefore conclude that

$$\lim_{\delta \to 0} x_\delta = x_{\text{MEY}} \qquad (2.67)$$

where x_{MEY} (Maximum Economic Yield) is the biomass level that maximizes sustained net revenue R.

In general x_δ is a decreasing function of δ, and we thus have

$$x_{\text{OA}} < x_\delta < x_{\text{MEY}} \quad \text{for } \infty > \delta > 0 \qquad (2.68)$$

This says that the optimal biomass level x_δ is a compromise between the extreme cases of zero and infinite discounting. Figure 2.10 displays this result graphically.

Present–Future Tradeoff

Before proceeding, let us use the whale example to further highlight the nature of the tradeoff between current and future harvesting. Imagine that the blue whale population is currently at its MEY level of $x_{\text{MEY}} = 76,000$ whales. Maintaining the population at this level will provide an annual catch of 1,875 whales, with a net sustained income of $ 12.9 million/yr. Compare this with the strategy of harvesting, say 10,000

Table 2.4. *Net annual revenue for two examples of blue-whale harvest strategies. Parameter values as in Table 2.2.*

Strategy	Net Annual Revenue (Millions of $/yr)	
	First Year	Subsequent Years
$x = 76,000$ MEY	12.78	12.78
$x = 66,000$	80.02	12.54

additional blue whales this year and subsequently maintaining a reduced population of 66,000 whales. Table 2.4 shows the results.

The first-year gain of $67 million is immense in comparison to the reduction in subsequent annual revenue, about $240,000/yr. Could the whaling companies have been expected to sacrifice a $67 million immediate income for a perpetual income supplement of $240,000 per annum? Not likely.

But there is another lesson to be learned from the example. How would the whalers know what to do?

What do the Fishermen Know?

The whalers presumably knew pretty well how much money they could expect to make from the next season's operations. They also probably had a rough idea of the extent to which blue whales were becoming more scarce. If the Schaefer logistic model were applicable, the blue whale population would be at its MSY level when the catch rate (i.e. catch per day of searching) had fallen to 50% of its original level. Should the whalers have agreed to maintain the population at a constant level from then on? How would they know what annual catch rate would achieve this result? Other than trial and error there is no way the whalers could know this. Calculation of sustainable yields requires sophisticated modeling and statistical analysis, based on the historical data.

Exactly the same situation applies to any developing fishery. The fishermen know where and how to catch fish, but they don't know how much can be caught sustainably. Nor do they usually have any way among themselves of controlling the total annual catch. All this requires a management authority with scientific expertise and the ability to control harvesting. It is not clear that the fishermen would be entirely happy with the managers' regulations, which will inevitably try to prevent the fishermen from catching fish that they know are available.

What do the Scientists Know?

This is not the place to enter into the arcane topics of modeling fish population dynamics and estimating population abundance. But let us imagine (as was the case for the whale population) that the fishery has kept careful track of its historical catches and effort levels. Let the scientists hypothesize a logistic growth model with parameters r and K, and a Schaefer catch model with catchability q. Various methods have been devised to estimate these parameters from the fishery data. (Usually a more complicated, age-structured model is used, if fishery data includes the age structure of annual catches.)

Typically the estimated parameter values involve substantial uncertainty. Likewise, stock estimates are also often highly uncertain. On top of this, the validity of model assumptions is often largely unknown. Regardless of all this uncertainty the scientists are charged with recommending an annual catch quota.

In the case of Antarctic blue whales, the population had been almost eliminated before the first scientific studies were undertaken in 1963. But the methods developed at that time were applied to fin, sei, and later Minke whales. Fin and sei whales were nevertheless overfished. By 1982 a general moratorium was placed on Antarctic whaling, and this was still in effect in 2005. Meanwhile the Scientific Committee of the IWC has met annually, and has developed a variety of potential management strategies for consideration if whaling ever resumes. Numerous papers have been published in the Reports of the International Whaling Commission.

In any event it is clear that scientific information is often much less complete than ideally would be desirable. Consequently there is often a bias towards demanding large catch quotas, on the grounds that the scientific evidence does not support the need for smaller quotas.

What Use is the Dynamic Optimization Model?

So does the dynamic optimization model of Eqs. (2.54)–(2.58) have any practical significance? Fishermen can't have the information needed to determine the optimal harvest strategy. Management agencies, who have the information, typically recommend MSY-oriented harvesting strategies with no economic content at all.

The discounting aspect of the optimization model suggests that fishermen and fishery managers may disagree about catch quotas, although other factors (such as overcapacity) may also be involved.

The optimization model indicates that the optimal biomass level x_δ may be well below x_{MEY}. As explained above, large profits can often be made by fishing down the stock, sometimes with an apparently small effect on long-term sustainable yield. Thus the perennial arguments between fishermen and fishery scientists over annual catch quotas may to a large extent be a consequence of time discounting, even if explicit quantitative dynamic optimization models are not used. The disagreement will be especially strong if the species is highly valued and easily captured. The more profitable a given fishery is, the more likely it is to be overexploited, even under management.

The critical question then becomes, is there some way to manage fisheries that would reverse this bias towards overfishing? We next turn briefly to this question (which is studied in detail in Chapter 4).

Economic Incentives

As noted above, conditions of unregulated open access in effect force fishermen to behave as if they had an infinite discount rate. The next question is whether various methods of fishery management would alter fishermen's behavior to more clearly resemble the economic optimum of finite discounting.

Consider the case of a fishery currently in a state of overexploitation, for example bionomic equilibrium. Would a TAC-based policy designed to facilitate stock recovery to the level of maximum sustained yield, change the fishermen's economic incentives so that they would support this policy? If not, why not?

There are two important reasons why this may not occur. First, annual catches will have to be reduced from their present level to allow for stock recovery. Though temporary, this recovery phase may be quite lengthy—years, or even decades in some cases. Fishermen's incomes will be reduced, at least initially. Fishermen who must make payments on their vessels may face bankruptcy.

But eventually, it might be argued, catches will increase and fishermen's incomes will grow accordingly. The investment of a few years' hard times will be repaid may times over by higher catches (and lower costs) in the future.

Now comes the second part of the argument. If indeed fishing incomes become positive, additional fishermen will be attracted to the fishery. Unless this new entry is prevented, rents will again be reduced to near-zero levels. The fishermen's long sacrifice will go for nought. Assuming that the original fishermen can predict this outcome, is it any wonder that they might fail to support the stock recovery program?

This argument can be pursued further (limited licensing, gear re-

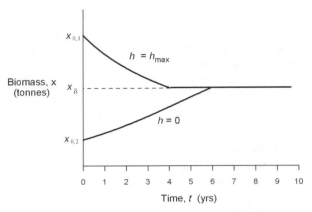

Figure 2.11 Optimal biomass trajectories $x(t)$ and harvest rates $h(t)$ for (1) an unexploited stock with $x_0 > x_\delta$, and (2) an overexploited stock with $x_0 < x_\delta$.

strictions, trip limits, and so on), but we postpone further discussion to Sec. 2.6. The main point is that management based on TACs alone cannot be expected to improve the economic performance of the fishery, and will often lead to an unwarranted increase in fishing capacity. Whether a combination of TACs plus limited vessel licensing will be likely to produce better results will be discussed later.

Recovery of a Depleted Fish Stock

Returning to our dynamic optimization model, what is the optimal fishing strategy when the current stock level $x = x(t)$ is different from the target biomass x_δ? As shown in Box 2.1, the optimal effort strategy (according to our current model) is

$$E^* = \begin{cases} E_{\max} & \text{if } x > x_\delta \\ G(x_\delta)/qx_\delta & \text{if } x = x_\delta \\ 0 & \text{if } x < x_\delta \end{cases} \tag{2.69}$$

This effort strategy will cause the stock $x(t)$ to move as rapidly as possible to the long-term equilibrium at x_δ (Figure 2.11). Such a strategy is sometimes called a "bang-bang" control strategy.

All this is very neat mathematically, but would the bang-bang effort strategy ever actually be used as a management strategy? Are there practical reasons, not included in the model, why some other strategy would be better? (One should always be suspicious of policy prescriptions that purport to be optimal. The question is, optimal under what assumptions?)

In fact, there are often good reasons why the bang-bang strategy may not be desirable. Here are two:

1. Harvesting a large stock at the maximum rate may depress the price of the product.
2. Placing a depleted stock under a moratorium may subject the fishing industry to severe hardships.

Our dynamic model can readily be altered to allow for the possibility that price depends on the quantity of fish brought to market. We can include this effect in our model by letting price $p = p(h)$ where h is the catch rate. We assume that $p(h)$ is decreasing in h (increased supply implies reduced price). Thus we now have

$$R = p(h)h - cE = p(h)h - c(x)h \qquad (2.70)$$

and the optimization objective becomes

$$\underset{\{h(t)\}}{\text{maximize}} \int_0^\infty e^{-\delta t} R \, dt \qquad (2.71)$$

with R as in Eq. (2.70).

To see how price sensitivity affects the optimal harvest strategy, we first ignore stock dynamics. Then, to maximize net revenue at any given time we require that $dR/dh = 0$, or

$$\frac{d}{dh}(p(h)h) = c(x) \qquad (2.72)$$

(This says that marginal revenue equals marginal cost, as in many microeconomic models.) As shown in Fig. 2.12, this equation implies that $h = h_1(x)$, where $h_1(x)$ is an increasing function of x.

The optimal harvest rate $h = h_1(x)$ therefore constitutes a feedback strategy for harvesting the population. This feedback strategy will drive the resource stock to an equilibrium level x^*, the long-term target biomass (Fig. 2.13). This idea is similar to the reference-point harvest strategies discussed in Sec. 2.1. In particular, the optimal harvest strategy is no longer the bang-bang strategy of Eq. (2.69). Rather, the harvest rate $h_1(x)$ is now adjusted gradually toward the optimal equilibrium $h^* = G(x^*)$, as the biomass $x(t)$ approaches x^*, instead of being switched off and on suddenly as $x(t)$ passes through x^*. Note, however,

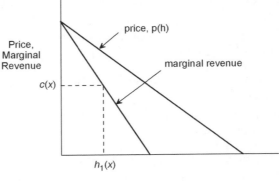

Figure 2.12 Price $p(h)$ and marginal revenue $(hp(h))'$ for the case that $p(h) = a - bh$. The harvest rate $h_1(x)$ that maximizes net revenue $p(h)h - c(x)h$ is an increasing function of stock size x (because $c(x)$ is decreasing in x).

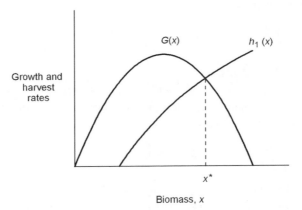

Figure 2.13 Feedback optimal harvest strategy $h_1(x)$ for the case of an elastic price $p = p(h)$. Biomass level x^* is the long-term optimal equilibrium stock level.

that the optimal harvest rate does become zero at sufficiently low stock levels.

The simplified treatment based on Eq. (2.72) is incorrect in two ways (see further discussion below), but the result in Figure 2.13 is qualitatively correct. The optimal harvest rule is a continuously increasing feedback strategy $h = h_1(x)$, as shown. Catch rates should be reduced if

$x(t)$ falls below the target biomass x^*, but not to zero unless $x(t) \ll x^*$. (The difficulties of implementing such a feedback strategy were discussed in Sec. 2.1.) Thus the fishing industry is not subject to sudden shut-down shocks whenever resource stocks decline marginally below the long-term target level. Rather, limited catch reductions are used, to allow for gradual stock recovery. Severe stock depletion, however, still calls for a harvesting moratorium.

Now, in what ways is the above analysis incorrect? First, in assuming that optimal fishing would maximize revenue $R = p(h)h - c(x)h$ we are taking a monopolistic view of optimal production. The expression $p(h)h$ for gross revenue from fishing includes monopoly profits. From a social equity standpoint this should be replaced by the social utility function $U(h) = \int p(h)\, dh$. Thus Eq. (2.72) should then be replaced by

$$p(h) = c(x) \tag{2.73}$$

and this implies a larger catch rate h than given by Eq. (2.72). For example, if $p(h) = a - bh$, we obtain

$$h_1(x) = \frac{a - c(x)}{2b} \quad \text{(Monopolistic case)}$$

$$h_1(x) = \frac{a - c(x)}{b} \quad \text{(Socially optimal case)}$$

Here the monopolist is a strong conservationist, holding back on production in order to increase his profits! (But note that both formulas agree qualitatively with Fig. 2.13.) The second shortcoming of the analysis is that it ignores the effects of harvesting on the fish population itself. Since harvesting negatively impacts the population, the optimal harvest rate is in fact smaller than that given by Eq. (2.73). This in turn implies a larger equilibrium biomass x^* than otherwise. For the socially optimal case of Eq. (2.73), x^* is given by

$$G'(x) - \frac{c'(x)G(x)}{p(h) - c(x)} = \delta; \quad h = G(x) \tag{2.74}$$

Note that this is in close agreement with our previous result for constant price, Eq. (2.63). Equation (2.74) can be rewritten as

$$p(h) = c(x) + \frac{c'(x)G(x)}{G'(x) - \delta} \tag{2.75}$$

The second term on the right can be shown to be positive, so we conclude that the solution h is smaller than that obtained from Eq. (2.73), at least at equilibrium. (For the derivation of these results see Clark and Munro 1975, or Clark 1990, Ch. 4.)

Once again, what is the practical significance of all this theory? No one expects a management agency to determine the target biomass level on the basis of such a complex, dynamic bioeconomic model. But since our basic bioeconomic model is essential for the management-related arguments to be made in the rest of this book, we need to be aware of the limitations of our main model.

But it is surely worthwhile knowing that under an incentive-based management system, the industry might actually press for *smaller* catch quotas than recommended by managers on purely biological grounds, in order to keep prices high. Also it is worth knowing that the extreme strategy of shutting down the fishery whenever stocks are below the target level, may be economically unwise, if small catches can be sold at a high price. Allowing a limited fishery on a depleted stock may be defensible on economic grounds, provided that the fishery can be rigorously monitored and controlled.

Irreversible Changes

An underlying assumption of the logistic growth model, Eq. (2.3), is that the natural growth rate $G(x)$ is positive for all stock levels, whenever the stock is not being harvested. In other words, any level of depletion short of actual extinction is assumed to be reversible over time, under a fishing moratorium. But as noted in Sec. 2.2, some populations may have depensatory dynamics, with a critical population level x_{crit} such that $G(x) < 0$ for $x < x_{\text{crit}}$. Our model then predicts ultimate extinction if $x(t)$ is ever reduced below x_{crit}. In practice, the value of x_{crit} will seldom if ever be known in advance, but supposing that it were known, would our bioeconomic theory ever predict that deliberate extinction is economically "optimal"?

First, unregulated open access fishing will reduce the population to a sub-critical level if and only if $x_{\text{OA}} < x_{\text{crit}}$. We also know that $x_\delta \to x_{\text{OA}}$ as $\delta \to \infty$. The following two conditions are therefore necessary and sufficient for deliberate extermination of a population to be theoretically optimal:

$$(a)\ x_{\text{OA}} < x_{\text{crit}}$$
$$(b)\ \delta \text{ is sufficiently large}$$

But would deliberate extermination of a valuable renewable resource ever be really desirable, even if a great deal of money could be made in the process? Surely not: our valuable marine resources ought to be preserved for future generations regardless of any profit-maximizing motives that might exist. Overfishing to the point of collapse, where it has occurred, has to be considered as an anomaly. Most such examples have doubtlessly been inadvertent rather than deliberate. The lesson to be learned is that marine populations may be more subject to irreversible overfishing than previously recognized. A more cautious, risk-averse approach to management appears to be developing in response to past failures (Chapter 5). If this is combined with strongly incentive-based methods of management (Chapter 4), the future of marine fishing may be bright.

Stock Assessment

Implementation of any given feedback harvest strategy requires that the current stock level be known with some accuracy.

We do not discuss the details of stock assessment in this book, but a brief list of methods is:

1. CPUE method;
2. Cohort analysis and related methods;
3. Sampling surveys;
4. Tagging experiments.

The first two methods use fishery-generated data, while the second two are based on research data that are independent of the fishery. For detailed discussion see for example Gulland (1983), Hilborn and Walters (1992), Punt and Hilborn (1997), Quinn and Deriso (1999), Walters and Martell (2004).

Various problems that are recognized in stock assessment include:

(a) bias, i.e., a given method may tend to systematically overestimate (or underestimate) stock abundance as the fishery develops;
(b) wide confidence intervals, i.e., numerical estimates of abundance are subject to large uncertainty;
(c) unreported catches, e.g., by-catches, discards and illegal catches, that can distort the estimates;
(d) high cost, e.g., for annual research surveys.

One viewpoint of the limitations of stock assessment is that of Walters and Martell (2002): "... existing assessment methods have often failed, and will probably continue to do so in the future despite more elaborate estimation methods." Similarly, in a recent, thorough review, Quinn

(2003, p. 381) states that "...one simulation exercise from NRC shows that the most recent exploitable biomasses are overestimated by all assessments, which include state-of-the-art techniques."

Yet reliable stock assessments appear to be essential for any system of sustainable fishing. How is this dilemma to be resolved? Let me mention two possible approaches. The first is input (effort) controls rather than output (catch) controls. The idea is that if the catch equation $h = qEx$ is valid, then controlling effort E will automatically reduce the harvest h at low stock levels x. I believe that the discussion of CPUE profiles in Sec. 2.2 is enough to throw serious doubt on the long-term validity of this approach. To reiterate, the Schaefer model $h = qEx$ is probably seriously flawed for most fisheries. In addition, controlling real effort (in contrast to nominal effort) over the long run is difficult in practice. Indeed, even if effort quotas are individualized (e.g., so many days fishing per year per vessel), this method inevitably encourages overfishing and capital stuffing (see Chapter 4).

As argued in Chapter 4, individual catch quotas (ICQs) can encourage conservation-oriented behavior patterns by fishermen, at least if the quotas are transferable. This advantage of individual quotas is less pronounced under individual effort quotas.

A second approach to overcoming the limitations of stock assessment revolves around risk management (Chapter 5). Under this approach, a substantial segment of each fish stock would be entirely protected from fishing. The principal method for this is based on marine protected areas (MPAs), or closed areas, in which all fishing is prohibited. Although a large literature now exists on this topic, very few discussions pertain to the risk-management aspect of MPAs. See Chapter 5 for further discussion.

2.5 Uncertainty

The above discussion emphasizes the role of large levels of uncertainty about the dynamics of exploited marine populations. Critical threshold stock levels cannot be predicted in advance, and unpredictable environmental fluctuations may affect all aspects of population dynamics, in unknown ways. In addition, stock estimates may be quite inaccurate, catch and effort data may be unreliable, and model parameters used to calculate allowable catch quotas may be highly uncertain and inaccurate.

What are the economic consequences of these uncertainties? The all

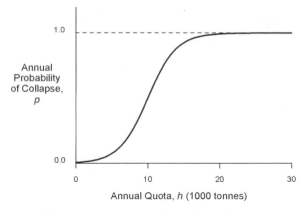

Figure 2.14 Annual probability of collapse $p(h)$, given an annual quota h. The functional form used here is $p(h) = 0.5\{1 + (1 - e^{-k(h-h_0)}/(1 + e^{-k(h-h_0)})\}$ where $h_0 = 10$ thousand tonnes/yr, $k = 0.5$. The maximum possible yearly catch is $h_{\max} = 30$ thousand tons, which wipes out the stock in one year, with very high probability.

but universal view of the fishing industry has been that scientific uncertainty means that catch restrictions are not justified. Future catches may admittedly be uncertain, but if the fish are out there today, why not catch them now? As noted previously, this notion is consistent with time discounting, which is increased by future uncertainty.

The following simple model partially captures these ideas. Suppose that a management agency wishes to set an annual catch quota h for all future catches. If h is too large, the fishery risks collapse. Let $p = p(h)$ be the probability that the fishery will collapse in any given year, given h. I imagine that the graph of $p(h)$ is as shown in Fig. 2.14.

For simplicity, assume that annual revenues are proportional to h. The expected present value of all catches is then equal to

$$EPV = \left\{1 + \frac{(1-p)}{1+i} + \frac{(1-p)^2}{(1+i)^2} + \ldots\right\} h \qquad (2.76)$$

$$= \frac{(1+i)h}{p(h) + i} \qquad (2.77)$$

where i is the annual discount rate. (The annual discount rate i is related to the instantaneous rate δ used previously, by the equation $1 + i = e^{\delta}$, or $\delta = \ln(1+i)$.)

The fishing industry would like a quota h that maximizes the value

Table 2.5. *Optimal annual catch quota h_i, and resulting annual probability of collapse p, for the model of Eq. (2.77).*

Annual discount rate i	Optimal catch quota, h_i (tonnes)	Annual probability of collapse, p
1%	2,800	2.7%
3%	3,700	4.1%
5%	4,200	5.2%
7%	4,500	6.0%
10%	5,000	7.6%
$\geq 11\%$	30,000	$> 99.9\%$

of the expression in Eq. (2.77). For the example shown in Fig. 2.14, the optimal quota h_i is given by Table 2.5.

As the discount rate i is increased, larger quotas become desirable, and the accepted risk of collapse increases. At $i = 11\%$ per annum the optimal quota jumps to $h_{max} = 30,000$ tonnes, a catch that all but ensures the immediate destruction of the stock. This prediction was a surprise to me, but it does make sense. The quota of 5,000 tonnes (optimal for $i = 10\%$) is already somewhat risky. At any larger discount rate, the "sure thing" harvest of 30,000 tonnes is the most profitable strategy.

(The numbers in Table 2.5 depend on model parameters, particularly the shape parameter k—see Fig. 2.14. For large k, the model approaches the deterministic case, with $p(h) = 0$ for $h < 10,000$ and $p(h) = 1$ for $h > 10,000$. Thus $h_{MSY} = 10,000$ tonnes, and this is superior to immediate extermination ($h = 30,000$) only if $i < 33.3\%$. For sufficiently small k, all harvest levels are highly risky, and $h = 30,000$ is optimal at all discount rates.)

This model indicates that deliberate overfishing to the point of extinction may be encouraged by uncertainty over future availability of the resource. Such uncertainty accumulates over time, so its economic effect is similar to that of discounting. The above model has the realistic twist that harvesting the resource increases the risk of collapse. This results in a positive feedback between financial discounting and uncertainty discounting.

The phenomenon of uncertainty discounting presents a dilemma to fishery managers. In negotiations with the fishermen, should managers admit to the full level of scientific uncertainty, or should they conceal this uncertainty on the grounds that it would only encourage demands for high quotas? Proponents of the precautionary principle argue that uncertainty should lead to reduced quotas (relative to some computation based on averages) in order to protect against stock collapse. We discuss these matters more fully in Chapter 5.

2.6 Management Implications

This chapter demonstrates that we cannot hope to comprehend the phenomenon of persistent overfishing unless we carefully examine the economic incentives of the fishing industry. Important as it may be, open-access competition is not the whole story. In many situations, fishermen may actually prefer overfishing to maintaining a sustainable fishery. Sacrificing short-term gains to achieve sustained long-term catches may not appear desirable to the fishermen, for several reasons, including discounting and future uncertainty. However, perhaps the most important factor underlying anti-conservationist behavior is the lack of ownership, or property rights (for want of a better phrase) in an open-access fishery. Fishermen might favor management strategies that would increase future catches, even at the cost of a short-term reduction in catches. But if the future gains will have to be shared with new participants in the fishery, today's fishermen will receive few of the future benefits. To state the matter differently, one does not wish to make an investment unless there are good prospects of receiving future rewards as result. How individual catch quotas can alter fishermen's economic incentives so as to favor investing in fish stocks is discussed in Chapter 4.

TAC-Regulated Management: The "Derby" Fishery
Many marine fisheries have been managed primarily by means of TACs—Total Allowable Annual Catch quotas. The TAC is calculated from a model of the population processes of growth, natural mortality, fishing mortality and recruitment (Sec. 2.7). Typically the TAC is designed to achieve maximum sustainable yield, with necessary adjustments in the event that the current stock is below the estimated MSY level.

Regardless of what additional management measures may be imposed, TAC-based catch controls seem unavoidable if sustained catches are to be maintained. An all-but universally encountered phenomenon for fisheries managed by TACs alone, however, has been the occurrence of a "derby"-type fishery, in which participants compete vigorously to obtain a share of the quota before the fishery is closed for the remainder of the season

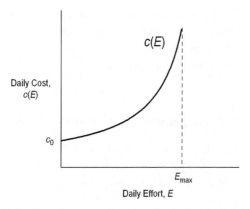

Figure 2.15 Daily effort cost function for individual fishing vessels. Net daily revenue equals $pqxE - c(E)$: see text.

(this being the usual method of achieving the TAC). Any fisherman who fails to do everything possible, consistent with his own economic interests, to grab an early share of the TAC, will be left behind.

There are two types of actions that a fisherman can take in a derby fishery. First, he can exert a large daily effort while the season remains open. Second, he can upgrade his equipment. Both actions are costly, so the fisherman faces a standard type of optimization calculation.

Let us consider the first possibility. Let E denote the fisherman's daily effort and let $c(E)$ be his daily cost. As before (see Eq. 2.45) we assume that

$$c'(E) > 0 \quad \text{and} \quad c''(E) \geq 0 \quad \text{for } 0 \leq E \leq E_{\max} \qquad (2.78)$$

as shown in Fig. 2.15. This is a realistic assumption, since costs are likely to escalate rapidly if daily effort is high. Here $E_{\max}$ denotes the vessel's maximum daily effort. The cost function $c(E)$, and the maximum $E_{\max}$ are determined by the particular configuration of vessel and gear that the fisherman possesses. By making capital improvements to his vessel or gear the fisherman may be able to reduce costs $c(E)$ and increase effort capacity $E_{\max}$, but we ignore this possibility for the moment.

The value c_0 given by

$$c_0 = c(0+) \qquad (2.79)$$

is the daily mobilization cost, i.e., the minimum expenditure required to go fishing at all on a given day. We naturally assume that $c_0 \geq 0$; in fact we will usually assume that $c_0 > 0$. One simplifying assumption

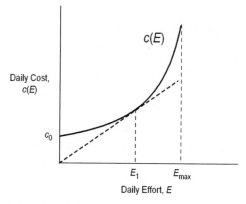

Figure 2.16 Graphical determination of effort level E_1 defined by Eq. (2.81); a tangent line from the origin to the cost curve meets that curve at $E = E_1$.

will be that all fishermen (more accurately, all vessel owners) have identical effort–cost functions $c(E)$. This is unrealistic, of course, but the alternative makes for a more complicated theory (Clark 1990, Sec. 8.1).

Whenever the fishery is open, the individual fisherman is assumed to maximize his daily net revenue:

$$\text{maximize } R = pqxE - c(E) \tag{2.80}$$

Define the effort level E_1 by the condition

$$c'(E_1) = c(E_1)/E_1 \tag{2.81}$$

Figure 2.16 shows a geometric construction for finding E_1. Note that $E_1 = 0$ if $c_0 = 0$, but otherwise $E_1 > 0$. The effort level E that maximizes daily net revenue R is given by

$$\left.\begin{array}{lll} E = 0 & \text{if} & pqx < c'(E_1) \\ c'(E) = pqx & \text{if} & c'(E_1) \leq pqx \leq c'(E_{\max}) \\ E = E_{\max} & \text{if} & pqx > c'(E_{\max}) \end{array}\right\} \tag{2.82}$$

(This is the same as Eq. 2.48, but with simplified notation. The previous argument is repeated here, for convenience.) This can be explained by inspection of Fig. 2.16. First, if $pqx < c'(E_1)$ then the line $pqxE$ lies below $c(E)$ for all E. Here $R < 0$ for all $E > 0$, so the optimal effort level is $E = 0$ (which avoids any cost). The other two cases in Eq. (2.82) follow in a similar way.

In particular, the fisherman does not go fishing at all if the stock level is too low, i.e. if $pqx < c'(E_1)$. Bionomic equilibrium in the unregulated open-access fishery is therefore given by

$$x_{\text{OA}} = \frac{c'(E_1)}{pq} \tag{2.83}$$

This agrees with our earlier characterization $x_{\text{OA}} = c/pq$, in the case of linear cost, $c(E) = cE$.

However, there is an interesting new prediction for the nonlinear case, if $E_1 > 0$. Namely, we can now predict the number of vessels N_{OA} that will operate at unregulated bionomic equilibrium. The total daily catch rate is $h = qx_{\text{OA}}N_{\text{OA}}E_1$ and this must equal the daily biological production rate $G_d(x_{\text{OA}})$. Thus

$$N_{\text{OA}} = \frac{G_d(x_{\text{OA}})}{qx_{\text{OA}}E_1} \tag{2.84}$$

We again predict complete rent dissipation at bionomic equilibrium:

$$R_{\text{OA}} = pqx_{\text{OA}}E_1 - c(E_1) = 0$$

by Eqs. (2.83) and (2.81) (Our theory of unregulated open access has become a bit too tidy at this point. It says that if a single vessel more than N_{OA} enters the fishery, then x falls below x_{OA}, so all vessels quit fishing! I leave it to the reader to explore the deficiencies of the present model in this respect.)

TAC-Based Regulation

We know from our earlier discussions that bionomic equilibrium is an undesirable outcome, involving minimal economic rents, and often resulting in severe overfishing and stock depletion. The purpose of TAC-type regulation is to prevent overfishing by limiting the annual catch to a sustainable level, for example MSY.

We therefore now consider a fishery that is managed using TACs, so as to maintain a certain target biomass x_{opt}. No other regulations are imposed; the TAC is achieved through seasonal closures of the fishery. What does our present model predict, in this case? Will the result be economically optimal, or if not, why not?

We assume that

$$x_{\text{opt}} > x_{\text{OA}}$$

Then, with $x = x_{\text{opt}}$ Eq. (2.82) implies that $E = E_0$ where

$$c'(E_0) = pqx_{\text{opt}} \quad \text{if} \quad pqx_{\text{opt}} < c'(E_{\text{max}})$$
$$E_0 = E_{\text{max}} \qquad \text{otherwise}$$

and we have $E_0 > E_1$. Daily net revenue per vessel is now positive:

$$R = pqx_{\text{opt}}E_0 - c(E_0) > 0$$

If N vessels participate in the fishery, the fleet's daily harvest rate is $Nqx_{\text{opt}}E_0$, and the total season's catch is

$$\text{Annual catch} = Nqx_{\text{opt}}E_0T$$

where T is the length of the seasonal opening, in days. (For simplicity we here ignore seasonal changes in the biomass x; the model can be adjusted to include this detail.) Thus, given an annual catch quota Q, we have

$$T = \frac{Q}{Nqx_{\text{opt}}E_0} \tag{2.85}$$

(unless this expression is $> T_{\text{max}}$, in which case $T = T_{\text{max}}$ and the full quota is not taken).

If $x_{\text{opt}} \gg x_{\text{OA}}$ then daily net revenue R under TAC-based management will be large. A large number N of vessels may then be attracted to the fishery, requiring drastic shortening of the fishing season. Examples in which the seasonal opening T has been reduced to a few days per year are commonplace.

Note that we cannot predict the exact number of vessels N that will participate in the TAC-regulated fishery, from the above model alone. To do so, we would need to consider also the fixed cost per vessel. We discuss this point briefly later (see also Chapter 3).

Next we consider the government's (or social) optimization problem for the TAC-regulated fishery. We assume first that a given number N of vessels participate in the fishery.

The problem is to maximize total sustained annual rent (we here ignore transitional dynamics, which are studied in Chapter 3):

$$\underset{E,T}{\text{maximize}} \, NT(pqxE - c(E)) \tag{2.86}$$

$$\text{subject to } H = NqxET \le Q \tag{2.87}$$

$$\text{and } 0 \le E \le E_{\text{max}}, \quad 0 \le T \le T_{\text{max}} \tag{2.88}$$

where T_{max} denotes the maximum possible season length, and where $x =$

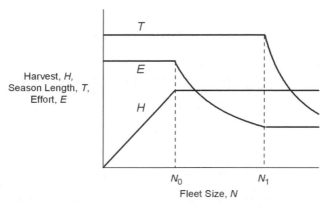

Figure 2.17 Optimal annual harvest H, season length T and daily effort per vessel E as functions of fleet size N, in a TAC-regulated fishery at equilibrium.

x_{opt}. The method of solution for this constrained optimization problem is outlined in Box 2.2. There are three cases, listed in Table 2.6. The fleet sizes N_0, N_1 are specified in Box 2.2. The optimal solution of Table 2.6 is also shown graphically in Fig. 2.17.

For example, if fleet size is small ($N \leq N_0$) then $T = T_{\max}$ and $c'(E) = pqx$. All vessels fish for the whole year, and maximize their daily net revenue. If a larger fleet exists, then individual effort E should be reduced towards E_1. If the fleet is very large ($N > N_1$) then $E = E_1$ and the season must be shortened.

How does the derby fishery compare with the optimum? We recall that for the derby fishery we have

$$E = E_0 \qquad \text{for all } N$$
$$H = \min(NqxE_0T_{\max}, Q)$$
$$T = \min(T_{\max}, Q/NqxE_0)$$

(where $x = x_{\mathrm{opt}}$).

The annual harvest H is the same for both cases, but if $N > N_0$ daily effort E is too high, and the season length is unduly short, in the derby fishery. This excessive effort and shortened season are what characterize the derby fishery.

Figure 2.18 shows the total fleet annual rents R for the two cases. The dashed line in the figure represents amortized annual fixed cost of the fleet. If entry to the fishery is unrestricted, fleet size will attain an

Box 2.2 Optimal fishing with TAC regulation

We write total annual net revenue as

$$J = J(E, T) = NT(pqxE - c(E))$$

Our optimization problem is

$$\underset{E,T}{\text{maximize}}\, J(E, T)$$

$$\text{subject to } H = NqxET \leq Q$$

$$\text{and } 0 \leq E \leq E_{\max}, \quad 0 \leq T \leq T_{\max}$$

There are two cases.
Case A: $NqxET < Q$, i.e., the quota is not taken.
In this case we have $\partial J/\partial E = 0$ (or $E = E_{\max}$). Define E_0 as the solution of this equation, so that

$$c'(E_0) = pqx \qquad \text{if this value } E_0 < E_{\max}$$

and

$$E_0 = E_{\max} \qquad \text{otherwise}$$

Now $J(E_0, T)$ is increasing in T. The constrained maximum must therefore occur at $T = T_{\max}$. Thus we have

$$NqxE_0T_{\max} < Q$$

and this will occur if N is sufficiently small, namely if

$$N < N_0 = Q/qxE_0T_{\max}$$

Note that N_0 is the fleet size that is just sufficient to catch the annual quota Q over the full season, using $E = E_0$.
Case B: $NqxET = Q$, i.e., the quota is taken. Here the optimization problem reduces to

$$\underset{E}{\text{maximize}}\, Q\left(p - \frac{c(E)}{qxE}\right)$$

$$\text{subject to } 0 \leq E \leq E_{\max}, \quad 0 \leq T \leq T_{\max}$$

Box 2.2. *continued*

where $T = Q/NqxE$. Maximizing the given expression is equivalent
to minimizing $c(E)/E$. If the constraints are not binding, then
$E = E_1$ where

$$c'(E_1) = c(E_1)/E_1$$

If the constraints are binding then we must still minimize $c(E)/E$
subject to the constraints.

Defining N_1 by

$$N_1 = Q/qxE_1T_{\max}$$

we see that (since $E_1 < E_0$)

$$N_0 < N_1 \leq +\infty$$

(Note that $E_1 = 0$ if $c_0 = c(0) = 0$; in this case N_1 is infinite.) As
above, N_1 is the fleet size that can capture the quota Q over the
full season, using $E = E_1$. Piecing this information together yields
the result in the text, Table 2.6, and shown in Fig. 2.17.

Table 2.6. *Optimal solution for the TAC-regulated fishery.*

Fleet size N	Daily effort E	Season length T	Annual catch H
$N \leq N_0$	E_0	$T_{\max}$	$NqxE_0T_{\max}$
$N_0 \leq N \leq N_1$	$Q/NqxT_{\max}$	$T_{\max}$	Q
$N \geq N_1$	E_1	$Q/NqxE_1$	Q

equilibrium at N_{ROA} ("Regulated open-access") in the TAC-regulated
derby fishery.

The optimal fleet size N^* for an optimally operating fleet can also be
determined from Fig. 2.18 (N^* is not drawn in the figure, for simplicity).
We have

$$N_0 < N^* < N_1$$

At this optimum optimorum, individual vessels do not use excessive

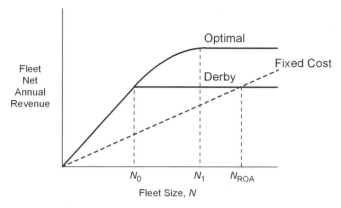

Figure 2.18 Fleet net annual revenue for a TAC-regulated fishery: (a) derby fishery; (b) optimal. Regulated open-access equilibrium fleet size is at N_{ROA}.

daily effort E_0 (as occurs in the derby fishery), so the optimal fleet size is larger than N_0.

We will not push the discussion further at this stage; many important additional details could be considered, including stock dynamics and adjustment, and fleet-size dynamics. These topics are taken up in Chapter 3.

The main predictions of the present model are that TAC-based fishery regulation, if it succeeds in preventing overfishing, will also preserve some (usually suboptimal) rents, but these rents will tend to attract an excess of fishing capacity. The ultimate result may be a "derby" fishery, characterized by a large fleet of vessels exerting high daily effort levels and capturing the annual TAC over a brief time span. (Note that, contrary to the usual verbal explanation of the derby fishery, it is not the imminence of the end of the season that motivates the fishermen's behavior, but simply their desire to maximize their daily net revenues.)

Trip Limits

TAC-based management often leads to a derby fishery, in which the entire annual quota is taken over a brief seasonal opening. In some cases the sudden supply of fish floods the market, driving down the price. By freezing the excess catch, the supply can be smoothed out over the year, but the frozen product may be less desirable to consumers.

One management response to this situation is the use of trip limits, whereby vessels are allowed a fixed quota per trip. Assuming that trips have specified maximum duration T_0, and a trip limit Q_0, the vessel owner's optimization problem becomes

$$\underset{E,T}{\text{maximize}} \ \ T(pqxE - c(E))$$

$$\text{subject to} \ \ H = qxET \leq Q_0$$

$$\text{and} \ \ \ T \leq T_0, E \leq E_{\max}$$

where the price p is now determined by supply and demand (but the individual vessel owner acts as a price taker, since his own catch has a minimal effect on the overall supply). Namely, the managers specify the period quota Q_w so as to maximize total revenue $p(Q_w)/Q_w$. Each vessel is then allocated the trip limit $Q_0 = Q_w/N$ where N is the number of vessels.

We can reasonably assume that the trip limits Q_0 will in fact be caught on each trip:

$$H = qxET = Q_0$$

The above maximization equation then becomes

$$\underset{E}{\text{maximize}} \, Q_0 \left(p - \frac{c(E)}{qxE} \right)$$

and this has the solution $E = E_1$, where $c'(E_1) = c(E_1)/E_1$, as previously. (If this implies that $T > T_0$ then the optimum becomes $T = T_0$ and $E = Q_0/qxT_0$.) The trip limit system thus also overcomes the problem of excess effort in the derby fishery. Trip limits are a sort of individual catch quota, applied on a weekly (for example) basis.

However, there remains the question of determining the fleet size N. Imagine, for example, that trip limits are introduced into an open-access, TAC-regulated fishery. The result will be an increase in annual revenues, and this will attract new entry to the fishery, leading to dissipation of the benefits of trip limits.

The next step, clearly, is to limit entry to the fishery. This is our next topic.

The Limited-Access Fishery

Under limited access, a fixed number of fishermen possess the right to exploit the resource, and all others are strictly excluded. The idea is that those who are licensed to exploit the resource will be motivated to conserve it in order to enhance their economic interest. Is this theory correct from the point of view expressed in the present chapter?

I trust it will come as no surprise that my answer is a qualified "maybe." Three essential qualifications are:

1. The exclusion of outsiders is strictly enforced. (However, license holders may be allowed to sell out if they wish, and this implies that outsiders can buy in.)
2. The harvesting activities of licensed individuals are carefully controlled.
3. Compensation by the license holders is made to the state, in return for the transfer of a public asset to private ownership.

Unless all three of these conditions are met, the limited-access system will not be both successful and fair. I will briefly discuss each of the qualifications.

First, exclusion of illegal participants is obviously essential, but is not necessarily a simple matter in the fisheries setting (nor, perhaps, for most other resources). How difficult it may be to exclude outsiders depends on the geographical distribution and mobility of the fish population. Transboundary and international fisheries are particularly problematic in this sense, although treaties may succeed in allocating access to the resource among participating states (see Munro 1987 for models). Also, within national territories, allocation of catches between various user groups, such as commercial, recreational, and aboriginal fishermen may be a necessary preliminary to the establishment of limited-access restricted rights within each group.

Assume now that N vessel owners have obtained fishing licenses. The fishery is managed using a TAC, but no other regulations are imposed. Will these owners be motivated to act in a responsible manner, refraining from overfishing or expanding the capacity of their vessels, while sharing the catch equitably among themselves? Here we will address this question intuitively; a rigorous game-theoretic model is discussed in Sec. 4.2.

Suppose that a certain licensed vessel has annual effort capacity $E_{\max}$, allowing an annual catch of $h = qxE_{\max}$, and an annual net revenue of

$$R = ph - cE_{\max} = (pqx - c)E_{\max}$$

The licensing authority has issued total vessel licenses such that the total catch $\sum h_i$ equals the TAC. We assume that the fishery is now profitable:

$$pqx - c > 0$$

The vessel's owner, fisherman Jones (say), contemplates making certain improvements to his vessel, which will increase its capacity from $E_{\max}$ to $E_{\max} + \Delta E$. The cost of these improvements is $c_1 \Delta E$. Jones's net annual revenue, adjusted for the amortized improvement cost, will be increased by the amount

$$\Delta R = (pqx - c - \delta c_1)\Delta E$$

where δ is the annual interest rate. Thus Jones will adopt the improvements provided that

$$pqx - c - \delta c_1 > 0$$

Each of the licensed fishermen will be motivated in the same way, with the result that fleet capacity will increase, if this inequality holds.

This prediction, of increased fishing capacity in a limited-entry fishery, has been experienced in virtually every known example. The management agency, as it monitors the fleet's catches, will observe that the TAC is being exceeded. This in turn will require shortening of the fishing season—another commonly experienced phenomenon in such management programs.

The conclusion from this model, that limited entry by itself will not usually prevent overcapacity from developing, is robust. The question as to what further management controls are required, will be studied in the next two chapters. The main conclusion will be that individual catch quotas are probably necessary, and sufficient, to prevent undue expansion of fishing capacity. When each fisherman has a fixed annual quota, the motivation for increasing his fishing capacity is removed. Experience with individual quota systems has shown that in many cases total fleet capacity is reduced once the quotas are in place. However, there may also be severe problems with such individual quota systems. Chapter 4 goes into the details.

The third criterion is one that is seldom discussed. At present the fishing industry in many countries receives vast subsidies. Some of these subsidies are direct payments, but others are indirect. Not the least of the indirect subsidies result from the failure of governments to charge appropriate royalties for the commercial use of a valuable publicly owned resource.

Does the state actually own the fishery resources within its 200-mile economic zone? This may be in doubt for species that move in and out of territorial waters, unless internationally shared ownership has been settled. However, once the state has established some form of limited entry in a given fishery, the implication is that the state does claim ownership of the resource (McRae and Munro 1989). If so, the right to exact royalties from users of the resource must exist (Macinko and Bromley 2002).

Almost by definition, an open-access resource is exploited with no royalties. However, once access to the resource becomes limited to certain individuals, social equity demands payment of appropriate user fees. Such fees, usually obtained by auction of harvest rights, are normal in forestry and elsewhere, though doubtlessly subject to political distortion. In fisheries, the distortion is often complete, with zero or near-zero royalties being charged in most cases. For example, in the USA, the 1976 Magnuson-Stevens Fishery Conservation and Management Act limited fees to 3% of ex-vessel value of landed fish, these fees being used to "recover the actual costs directly related to the management and enforcement of IFQ [Individual Fishing Quota] and CDQ [Community Development Quota] programs." (Ocean Studies Board 1999; this volume is a wide-ranging review of fisheries management at the US Federal level.)

But there is an important additional reason for collecting resource royalties in fisheries, namely the problem of anticipation. If fishermen (existing or potential) anticipate that entry to the fishery will be restricted in the future, there will be a strong incentive to expand the fleet now. Two reasons for this are, first that in most cases only fishermen with a history of participation in the fishery will be eligible for licenses, and second that a successful limited entry program will increase future profitability. Therefore the prospect of future limited entry may cause an anticipatory increase in fleet capacity, dissipating in advance much of the hoped-for benefits.

Resource royalties, if anticipated, would reduce or eliminate this incentive for fleet expansion. Substantial rent-capturing resource royalties are desirable for both equity and efficiency.

Several additional criteria for the successful management of fisheries could be added to the foregoing list. One important example is:

4. Management of the resource must explicitly take account of risk and uncertainty.

The topic of risk management is taken up in Chapter 5.

2.7 Age-Structured Population Models

The discussion of fishery bioeconomics in this chapter has been based entirely on the dynamic, so-called general production model

$$\frac{dx}{dt} = G(x) - h$$

Here the fish population is described by a single variable $x(t)$, interpreted as the biomass of the population at time t. The actual population processes of growth, reproduction and mortality are subsumed by the single function $G(x)$, and fishing mortality by the harvest rate h.

Actual fish populations have a much more complex structure than this. This structure includes: spatial distribution and migration; age, size and growth of individual fish; seasonal variation in growth, reproduction and mortality; characteristics of the marine environment, both physical and biological. In addition, all changes to population structure are subject to random variation over time; this is especially true of annual recruitment, which may vary over orders of magnitude for some species.

A full bioeconomic theory of fisheries would include these features, as well as a large number of economic complexities. But this theory would be immeasurably complex. In this section we will discuss an age-structured model of population dynamics that is often used in fisheries management. We draw certain bioeconomic predictions from this model, noting that these are qualitatively similar to the predictions based on the general-production model.

Growth Overfishing

In the Gordon–Schaefer model, overfishing occurs by definition if the biomass x falls below the level x_{MSY} that maximizes net production $G(x)$. This model does not specify exactly which population processes are affected by overfishing. On one hand, fishing may reduce the number of fish remaining in the population, thereby reducing the production of new, young fish. If excessive, this is called recruitment overfishing. On the other hand, fishing may remove most of the older fish, reducing the average age and size of remaining fish. This outcome, called growth overfishing, can have two effects. First, fish may be caught before they have grown to an optimal size, and second, recruitment may be reduced because older, larger fish that produce more offspring than young fish have been removed. To analyze these processes we need an age-structured model.

An Age-Structured Dynamic Model

To construct a dynamic age-structured fishery model, let $X_{a,t}$ denote the number of fish of age a, at the start of year t. We first consider the dynamics of $X_{a,t}$ within a single year:

$$\frac{dX_{a,t}}{ds} = \begin{cases} -(M_a + F_{a,t})X_{a,t} & 0 \le s \le s_1 \\ -M_a X_{a,t} & s_1 < s \le 1 \end{cases} \tag{2.89}$$

Here M_a denotes the natural mortality rate at age a and $F_{a,t}$ denotes fishing mortality rate at age a, year t. Also s_1 is the portion of the whole year during which fishing occurs. Solving Eq. (2.89) gives

$$X_{a,t} = \begin{cases} e^{-(M_a + F_{a,t})s} X_{a,t}(0) & 0 \le s \le s_1 \\ e^{-(M_a s + F_{a,t} s_1)} X_{a,t}(0) & s_1 < s \le 1 \end{cases}$$

We therefore conclude that (since $X_{a,t}(0) = X_{a,t}$ and $X_{a,t}(1) = X_{a+1,t+1}$)

$$X_{a+1,t+1} = e^{-(M_a + F_{a,t} s_1)} X_{a,t} \tag{2.90}$$

This is sometimes expressed as

$$X_{a+1,t+1} = e^{-Z_{a,t}} X_{a,t}, \quad Z_{a,t} = M_a + F_{a,t} s_1 \tag{2.91}$$

i.e., $Z_{a,t}$ is the total annual mortality for age a, which equals the sum of natural plus fishing mortality.

Next, let $Y_{a,t}(s)$ denote the cumulative catch (by weight) of age a fish in year t, for time 0 to s. Then

$$\frac{dY_{a,t}}{ds} = F_{a,t} X_{a,t}(s) W_a(s) \tag{2.92}$$

where $W_a(s)$ is the average weight of fish of age a, time s. If we make the simplifying assumption that $W_a(s) = W_a = $ constant, we can calculate the total annual catch from age a:

$$Y_{a,t} = Y_{a,t}(s_1) = \int_0^{s_1} F_{a,t} X_{a,t} e^{-(M_a + F_{a,t})s} W_a \, ds$$

$$= \frac{F_{a,t} X_{a,t} W_a}{M_a + F_{a,t}} (1 - e^{-(M_a + F_{a,t})s_1}) \tag{2.93}$$

This is called Baranov's catch equation, in honor of F.I. Baranov, a pioneer in fisheries science (Ricker 1954). Note that yield $Y_{a,t}$ is proportional to $X_{a,t}$, and also that $Y_{a,t}$ is an increasing function of $F_{a,t}$ with an asymptotic limit:

$$\lim_{F_{a,t} \to \infty} Y_{a,t} = X_{a,t} W_a \tag{2.94}$$

The next question concerns the relationship of age-specific fishing mortality to fishing effort E_t. Assuming a trawl-type fishery with Type II CPUE profile, we have

$$F_{a,t} = q_a E_t \tag{2.95}$$

where q_a is the catchability of age a fish. A special case, knife-edge selectivity, is characterized by

$$q_a = \begin{cases} 0 & a < a_1 \\ q & a \geq a_1 \end{cases} \tag{2.96}$$

where a_1 is the age of first capture.

The total annual catch is given by

$$Y_t(E_t) = \sum_{a \geq a_1} Y_{a,t} \tag{2.97}$$

From Eq. (2.93) we see that $Y_t(E_t)$ is an increasing function of E_t, with asymptotic limit:

$$\lim_{E_t \to \infty} Y_t(E_t) = \sum_{a \geq a_1} X_{a,t} W_a \tag{2.98}$$

High-effort fishing thus tends to remove all vulnerable cohorts from the population. After a sequence of such years, most of the surviving population consists of young fish of age $a \leq a_1$. The annual catch then consists almost entirely of fish of age a_1.

Stock-Recruitment

We next consider the question of reproduction and recruitment. We assume that the recruitment of age-one fish is a function of the previous year's stock level:

$$X_{1,t+1} = F(X_{1,t}, X_{2,t}, \ldots X_{A,t})$$

A commonly used example is due to Beverton and Holt (1957):

$$X_{1,t+1} = \frac{\alpha B_t}{1 + \beta B_t} \tag{2.99}$$

where B_t denotes egg production:

$$B_t = \sum_{a \geq a_0} f_a X_{a,t} \tag{2.100}$$

Here f_a is fecundity at age a, and a_0 is the age of sexual maturity. (We do not consider males and females separately, for simplicity.) In Eq. (2.99) α and β are constants.

Rewrite Eq. 2.90:

$$X_{a+1,t+1} = e^{-(M_a + F_{a,t}s_1)} X_{a,t} \quad (a \geq 1) \tag{2.101}$$

Equations (2.99)–(2.101) constitute our basic age-structured model of population dynamics. This model is standard in fisheries biology (e.g., Beverton and Holt 1957; Walters and Martell 2004).

Like the basic Gordon–Schaefer model, the foregoing model is deterministic. Most actual fish populations experience significant random effects, particularly in terms of recruitment. Thus we also consider a stochastic version of Eq. (2.99):

$$X_{1,t+1} = \frac{\alpha B_t}{1 + \beta B_t} \cdot \epsilon_t \tag{2.102}$$

where ϵ_t is a random variable (for example, lognormally distributed) with unit mean.

Unregulated Open Access

We next consider the case of unregulated open access. We first assume a given maximum effort capacity $E_{\max}$. Let $R_t(s)$ denote the flow of net fleet revenue at time s in year t. Then

$$R_t(s) = pY'_{t,s}(E_t) - cE_t$$

where $Y'_{t,s}(E_t) = dY_{t,s}/ds$ is the catch rate. We have

$$Y'_{t,s}(E_t) = \sum_{a \geq a_1} qX_{a,t}(s)W_a E_t(s)$$

Therefore (with p, c denoting price of fish and cost of effort)

$$R_t(s) = (pqB_t(s) - c)E_t(s) \tag{2.103}$$

where $B_t(s)$ is the fishable biomass

$$B_t(s) = \sum_{a \geq a_1} X_{a,t}(s)W_a \tag{2.104}$$

Under the open-access assumption, we have

$$E_t(s) = \begin{cases} E_{\max} & \text{if } pqB_t(s) > c \\ 0 & \text{otherwise} \end{cases} \tag{2.105}$$

Thus, depending on the maximum effort capacity $E_{\max}$ and the initial biomass $B_t(0)$, the open-access fishery will either persist for the entire feasible season $0 \leq s \leq s_1$, or will terminate early due to lack of fish. In a developing fishery, the annual initial biomass $B_t(0)$ ($t = 0, 1, 2, \ldots$) will typically decrease over time as the initial stocks are gradually fished down. Eventually an equilibrium will emerge, with $B_t(0) = B_{t+1}(0) = \cdots$ This is the bionomic equilibrium for our age-structured model. Annual rents are not zero at bionomic equilibrium, because each year's initial revenue flow $R_t(0)$ is positive. But because of uncontrolled overfishing these rents will typically be far smaller than they could be under proper management.

We have not yet discussed how the age at first capture a_1 is determined in an open-access fishery. In net-based fisheries (e.g., trawl, purse-seine, gillnet), a_1 is related to the mesh size used in the nets. Typically the fishermen will elect to use a minimum mesh size in order to maximize their catch rates, although this may be modified if sub-sized fish are of little value.

Fishery Management

Traditional approaches to managing age-structured fisheries are based on the concept of maximum sustainable yield, or a modification thereof. Suppose, for example, that the managers are able to control both fishing mortality F_t and the age of first capture $a_{1,t}$. Assuming that these are held constant over time ($F_t = F$ and $a_{1,t} = a_1$), we can use the above model to calculate a sustainable yield function $Y_{\text{sust}}(F, a_1)$. Optimal management is then defined in terms of maximizing this function. In practice, this objective is usually realized by regulating mesh size to achieve the optimal a_1, and regulating effort (through seasonal closures) to achieve the optimal F. As explained earlier, such a management strategy by itself often results in the overexpansion of fishing capacity, which must then be countered by further shortening of the fishing season.

A modification of this MSY-based approach that allows for a margin of error, uses the concept of F_x-fishing, defined as follows. First assume a pre-determined age of first capture. Sustainable yield $Y_{\text{sust}}(F, a_1)$ is then a function of F alone. For a given value of x ($0 < x < 1$), F_x is defined by the condition

$$Y'_{\text{sust}}(F_x, a_1) = xY'_{\text{sust}}(0+, a_1) \tag{2.106}$$

In other words, the slope of the sustainable-yield curve at $F = F_x$ equals x times the slope at the origin.

Clearly $F_x < F_{\text{MSY}}$, so that F_x-fishing is more conservative than

F_{MSY}, with a degree of caution that increases with x. A popular rule of thumb has been $F_{0.1}$.

Economically Optimal Fishing

Neither the MSY or F_x strategies consider economic aspects. If these are included, we can ask the following questions:

1. Should the specified effort level be reduced, relative to some value F_x, resulting in larger fish stocks, thereby lowering the cost of fishing?
2. Or should effort be increased to account for discounting?
3. In the case of a depleted stock, should rehabilitation occur rapidly (zero effort) or slowly (reduced effort)?

The same questions can be asked regarding the age of first capture.

Without going into details, we can state that the answers are analogous to those for the Gordon–Schaefer model. For example, cost effects imply lower effort levels and higher age of first capture, both of which would result in larger fish stocks and correspondingly lower costs of fishing. On the other hand, discounting implies higher effort levels and lower age of first capture, both of which trade off larger short-term economic benefits against smaller long-term benefits.

Are these theoretical results of any practical significance? I won't repeat the arguments that were given earlier, but it is worth emphasizing that the analysis in this chapter is based on deterministic models. Real fisheries are anything but deterministic. Today's crisis in fisheries is surely to some degree a result of failing to manage for risk. Having said so, however, it must be admitted that get-rich-quick incentives associated with high rates of discounting do seem to have had a dominant influence in marine fisheries. There's a lot of money to be made by removing valuable fish from the sea, and the devil take the hindmost. Any hope for reversing this trend must be based on a management system that somehow alters fishermen's incentives in a way that encourages conservation for long-term economic benefits. But this is the topic of Chapter 4.

2.8 Summary of Chapter 2

In this chapter we have presented our basic dynamic bio-economic model. Our presentation has two main thrusts: derivation and discussion of model predictions on the one hand, and the analysis of underlying assumptions and their implications on the other hand. These two types of analysis are essential for any theory that has practical applications.

The biological component of our basic model, called a general production model, is

$$\frac{dx}{dt} = G(x) - qEx \qquad (2S - 1)$$

(The meaning of the symbols is assumed to be familiar to the reader by now.) This model includes the following features:

- population dynamics
- effect of harvesting on the population
- relation between effort E and catch rate $h = qEx$

A common specification of the natural growth function $G(x)$ is the logistic form

$$G(x) = rx(1 - x/K) \qquad (2S - 2)$$

but the theory allows for a general form. The logistic model (like other similar models) assumes that population growth rate is positive for $0 < x < K$. This implies that the population will recover from any degree of depletion short of actual extinction. Recent empirical studies of some severely depleted fish populations indicate that this assumption may often not be justified.

The assumptions underlying the Schaefer catch equation

$$h = qEx \qquad (2S - 3)$$

were discussed in some detail. In particular, this equation implies that the catch-per-unit-effort (CPUE) statistic h/E is a direct index of current stock abundance x. It was argued that, although CPUE may reflect the average density $\rho(x)$ of the currently fished stock, in many cases this will not be proportional to x itself. In most fisheries CPUE tends to overestimate stock abundance at low stock levels. An alternative model that incorporates this feature is the power law

$$h = qEx^{\beta} \quad (0 < \beta < 1) \qquad (2S - 4)$$

Next, we introduced the economic parameters p (price of fish) and c (cost of effort), leading to the equation

$$R = ph - cE \qquad (2S - 5)$$

for the flow of net revenue to the fishing fleet. When combined with Eq. (2S − 3) this implies that

$$R = (pqx - c)E \qquad (2S - 6)$$

We then assumed that fishing would take place if and only if $R > 0$. More precisely, we supposed that for an unregulated, open-access fishery,

$$E(t) = \begin{cases} E_{\max} & \text{if } R > 0 \\ 0 & \text{if } R < 0 \end{cases} \qquad (2S - 7)$$

Here $E_{\max}$ denotes the maximum effort capacity, here assumed to be a given constant (related to fleet size); further discussion of $E_{\max}$ occurs in Chapter 3.

From Eqs. (2S − 6) and (2S − 7) we see that

$$E(t) = \begin{cases} E_{\max} & \text{if } x(t) > x_{\text{OA}} \\ 0 & \text{if } x(t) < x_{\text{OA}} \end{cases} \qquad (2S - 8)$$

where

$$x_{\text{OA}} = \frac{c}{pq} \qquad (2S - 9)$$

The time trajectory $x(t)$ of the harvested fish stock therefore satisfies

$$\lim_{t \to \infty} x(t) = x_{\text{OA}} \qquad (2S - 10)$$

(unless $E_{\max}$ is too small for this to happen—here we ignore this possibility). The stock level x_{OA} is referred to as the bionomic equilibrium of the unregulated open-access fishery.

In other words, effort $E(t)$ switches off in a knife-edge fashion when $x(t)$ hits x_{OA}. This prediction results from the tacit assumption that all fishing vessels have the same cost c and the same catchability q. If this is not the case then effort will gradually decline as high-cost vessels leave the fishery as the fish population is fished down. Figure 2.9 depicts the latter situation.

In the former case we have, from Eq. (2S − 6)

$$R = 0 \quad \text{when } x = x_{\text{OA}} \qquad (2S - 11)$$

This outcome is referred to as the (complete) dissipation of economic rent in the unregulated, open-access fishery. Note, however, that this prediction depends on the knife-edge assumption of Eq. (2S − 8). If this is not the case, then positive infra-marginal rents will persist at equilibrium, although usually these rents will be well below the maximum possible.

The prediction that an unregulated, open-access fishery will reach an equilibrium characterized by zero, or low, sustained economic yield can be interpreted in game-theoretic terms. Competitive fishing is a non-cooperative game with features similar to the famous prisoners' dilemma. Namely, although all participants would benefit from a unified reduction of effort, each would gain more by increasing his own effort. Likewise, if all participants exert excess effort, nothing can be gained by an individual who single-handedly reduces his effort. The outcome is a dynamic equilibrium with lower benefits than could be obtained—in theory at least—by using cooperative behavior. Much of the current crisis in world fisheries stems, we argue, from the use of management strategies that fail to address the economic realities of non-cooperative behavior in the exploitation of common-pool resources.

Next, an optimization, or cooperative, model is needed for comparison with the non-cooperative case. Our basic dynamic optimization model is specified by the equation

$$\underset{\{E(t)\}}{\text{maximize}} \, J = \int_0^\infty e^{-\delta t} R(t) \, dt \qquad (2\mathrm{S}-12)$$

We showed that this problem has the following solution:

$$E(t) = \begin{cases} E_{\max} & \text{if } x(t) > x_\delta \\ E_\delta & \text{if } x(t) = x_\delta \\ 0 & \text{if } x(t) < x_\delta \end{cases} \qquad (2\mathrm{S}-13)$$

where the target stock level x_δ is given by

$$G'(x_\delta) - \frac{c'(x_\delta)G(x_\delta)}{p - c(x_\delta)} = \delta \qquad (2\mathrm{S}-14)$$

Here $c(x) = c/qx$ and $E_\delta = G(x_\delta)/qx_\delta$. Thus the economically optimal harvest strategy *for our basic model* consists of adjusting the resource level $x(t)$ as quickly as possible towards the long-term optimal equilibrium $x = x_\delta$.

We showed that x_δ is a monotone decreasing function of the discount rate δ, and that

$$x_{\mathrm{OA}} < x_\delta < x_{\mathrm{MEY}} \qquad (2\mathrm{S}-15)$$

where x_{MEY} is the stock level that maximizes sustainable economic yield:

$$x_{\mathrm{MEY}} \text{ maximizes } R_{\mathrm{sust}} = (p - c(x))G(x) \qquad (2\mathrm{S}-16)$$

We also have

$$x_{\text{MEY}} = \lim_{\delta \to 0} x_\delta \quad \text{and} \quad x_{\text{OA}} = \lim_{\delta \to \infty} x_\delta \qquad (2S - 17)$$

i.e., x_{MEY} is the optimal target stock level under zero discounting, while x_{OA} corresponds to infinite discounting.

The optimal stock level x_δ usually differs from x_{MSY}, the stock level that provides the maximum sustained yield. It may happen that $x_\delta > x_{\text{MSY}}$, or $x_\delta < x_{\text{MSY}}$, depending on circumstances. On one hand, unit fishing costs $c(x)$ are decreasing in x, and this favors maintaining a standing stock larger than x_{MSY}. On the other hand, future discounting implies a preference for short-term gain (obtained by fishing down the stock) over long-term gains (from high sustained yield), so that discounting favors a standing stock smaller than x_{MSY}. Which effect dominates depends on all the parameters of the model.

Should these economic factors be incorporated into management policy? Perhaps so, but only if some method of preventing overfishing and rent dissipation has been put in place. See Chapter 4 for further discussion.

It is perhaps reassuring that the optimal fishing strategy leads to an equilibrium stock level and harvest rate, since virtually every actual management program has this intended characteristic (even if not based explicitly on economic considerations). However, it might be objected that the on-off, or "bang-bang" nature of the optimal effort strategy of Eq. (2S – 13) is unreasonable in both economic and practical terms. In fact, the on-off feature is a consequence of our basic model assumptions (specifically, the linearity of the control problem of Eq. (2S – 12) in the control variable $E(t)$) and this feature disappears if these assumptions are changed.

For example, if we assume that price is elastic, $p = p(h)$, or that effort costs are nonlinear, $c = c(E)$, then R becomes

$$R = p(h)h - c(E)$$

The result of these changes is that the on-off harvest strategy is replaced by a continuous feedback strategy of the form

$$h = h(x) \qquad (2S - 18)$$

Under such a strategy, harvests vary continuously as $x(t)$ approaches the target level. In particular, an overfished stock level $x < x_\delta$ does not necessarily imply that a fishing moratorium is optimal, unless the stock is severely depleted. Maintaining a small harvest rate can provide welcome revenues while also allowing for gradual stock recovery. (A different reason for the non-optimality of total fishing moratoria will be discussed in Chapter 3.) Also, with a continuous feedback strategy as in Eq. (2S – 18), random fluctuations of the fish population (our basic model does not allow for this) do not imply that the fishery will suffer from sudden openings and closures, unless the fluctuations are large.

An important practical requirement for any management strategy is the need for up-to-date stock assessments. Such assessments are often expensive and rather inaccurate, as expected for a resource that is both mobile and invisible. Various ways of minimizing the need for accurate stock assessments have been suggested, including effort quotas rather than catch quotas, and the use of closed areas to hedge against management error. The latter possibility will be discussed in Chapter 5.

Important aspects of fisheries not addressed in our basic model include

- population structure (age and size of fish, spatial structure, genetic diversity)
- natural fluctuations
- season effects (migration, feeding, spawning)
- ecosystem dynamics
- fixed costs and investment strategy
- uncertainty

Some of these topics are discussed later in the book.

A theoretical model of the interplay between uncertainty and discounting was discussed in this chapter. The model uses the idea that the risk of collapse of a harvested population may increase as the annual catch quota is increased. In spite of this risk, the optimal harvest strategy under future discounting may call for large catch quotas, the more so the higher the discount rate.

The question arises, what is the appropriate discount rate for government decisions in a renewable resource industry? We refer the reader to the extensive literature on this topic (see Sumaila and Walters 2005 for references); recommendations vary from zero discounting to ten per cent per annum (the official rate for the federal government of Canada). But regardless of what discount rate economists recommend, the preservation of valuable renewable resource stocks must surely be a top priority. Discounting is not an academic nicety; on the contrary it is possibly the most important parameter in the field of resource conservation. But conservation of biological resources overshadows discounting; if the official discount rate calls for overfishing of some stock, then the discount rate should be reduced, period.

The two models—unregulated open access, and dynamic optimization—indicate that substantial economic gains could be achieved through fishery management. What form should such management assume? The traditional approach has been based almost exclusively on population biology, with MSY as the leading paradigm. Scientists are appointed to estimate an annual catch quota (TAC), chosen to be as large as possible subject to the need to protect the resource. A seasonal opening is then determined so that the existing fleet will capture the TAC in the available time span. This method typically induces an expansion of fishing capacity (as explained in Chapter 3), leading to a progressive shortening of the fishing season. This in turn results in a "derby" fishery, in which the vessels compete vigorously to obtain a share of the catch during a brief seasonal opening. (In the case of whaling, the competition was sometimes referred to as the "whaling Olympics.") We presented a model of the derby fishery, and again compared it to the results of an economic optimization model.

One method that is sometimes used to control a derby fishery employs trip limits, which specify an individual catch quota for each fishing trip (of some specified duration). When combined with limited vessel licensing, trip limits are the same as individual catch quotas (IQs). Our model of the derby fishery indicates that the economic optimum can be achieved in this manner. Further discussion of individual quotas occurs in Chapter 4.

The general-production model of Eq. $(2S - 1)$ overlooks the age structure of the population. We therefore reviewed briefly the standard age-structured fishery model, sometimes called the dynamic-pool model (for some obscure reason). Optimizing the yield from an age-structured population involves controlling both the number of fish harvested per year, and the minimal age (or size) at which fish are taken. Under open-access conditions both of these aspects will occur sub-optimally. Overfishing will deplete the breeding stock (this is called recruitment overfishing), and also capture fish at too early an age (growth overfishing). In fact, growth overfishing also implies recruitment overfishing, because older, larger fish have greater fecundity than young fish.

The term "trophic overfishing" refers to the progressive depletion of large-bodied species from a given fishing area. These species, which are typically high on the food chain, are gradually replaced by smaller fish (Pauly et al. 1998; Myers and Worm 2003).

We fell short of analyzing a full dynamic economic optimization model of an age-structured fishery, as the analytic solution of such an optimization model does not seem feasible. However, for specific cases a simulation model can readily be devised for this purpose. Not surprisingly, such models often predict a strong sensitivity of the optimal harvest strategy to the discount rate.

3

Investment and Overcapacity

In this chapter we model the investment decisions of fishermen, under various management conditions. This study, which extends the results of Chapter 2, is necessary in order to understand the phenomenon of overcapacity, now widely considered to be a major obstacle to sustainable fisheries management. For example, Gréboval and Munro (1999), quoting P. Mace, state that "The key problem afflicting marine capture fishery resources ... is overcapacity .... Over the two decades 1970–1990, world industrial fisheries harvesting capacity grew at a rate eight times greater than the rate of growth of landings from world capture fisheries." And, according to Hilborn (2002) "Overfishing is primarily a symptom of overcapitalization and fisheries management systems that do not work." Many other similar statements could be cited.

At least three questions arise from these perceptions:

1. Why has overcapacity occurred?
2. In what ways is overcapacity undesirable?
3. How should overcapacity be dealt with?

Governments have already spent hundreds of millions of US dollars attempting to reduce excess capacity in fisheries by means of buy-back programs. Whether these programs have been successful in achieving their objectives is questionable, to say the least. In a detailed empirical study of buy-back programs, Holland et al. (1999) stated that most buy-back programs had failed, and that "... the potential for buy-back programs to achieve the ... goals that they are typically meant to address seem [sic] very limited." This raises a fourth question:

4. Can buy-back programs be expected to be successful in reducing over-capacity, or if not, why not?

The models to be discussed in this Chapter will strongly suggest that buy-back programs are unlikely to succeed over the long run. In fact, our theory will indicate that buy-back programs may actually make matters worse if fishermen anticipate that buy-backs will occur in the future. We will also argue that if buy-backs are voluntary, the cost of a successful buy-back program may be astronomical.

3.1 Fixed versus Variable Costs

First we recall the way that costs of fishing were included in our dynamic model of Chapter 2, namely

$$\text{Cost of effort} = cE \tag{3.1}$$

where c is a constant cost coefficient. Let us think about costs in a bit more detail, however. Imagine a trawl fishery, carried out by a fleet of identical trawlers. One trawler is taken as the unit of effort, so we can define $E(t)$ = total number of trawlers active at time t. Variable costs include costs of fuel and oil, food and supplies, vessel and gear maintenance, and labor. It is probably reasonable to assume that these costs are incurred only when a vessel is actively trawling, so that Eq. (3.1) is realistic, at least for total fleet variable costs.

These costs, however, do not include capital costs of the vessel and its gear. The latter are fixed costs that the vessel owner pays once, upon purchase of the vessel. All models discussed so far have ignored fixed costs. This is equally true for the static model of Chapter 1 and the dynamic models of Chapter 2. Capital costs of vessels and gear are an important component of fishermen's economic decisions, however. Unless a potential vessel owner can expect to recover these capital costs from fishing, he will not wish to enter the fishery. On the other hand, once a vessel-cum-gear has been purchased, the owner will tend to continue fishing as long as his revenues exceed his variable costs. In this chapter our dynamic models are extended to include these aspects.

3.2 The Open-Access Fishery

We begin with the case of the unregulated open-access fishery, based on the biological model

$$\frac{dx}{dt} = G(x) - qEx, \quad x(0) = x_0 \tag{3.2}$$

Effort $E = E(t)$ is constrained by

$$0 \le E(t) \le E_{\max} \tag{3.3}$$

where $E_{\max}$ denotes total fleet size, or more accurately, total fleet effort capacity. We now wish to predict both the total fleet size $E_{\max}$ and the time schedule of effort $E(t)$ under open-access conditions. That is, we assume that fishing activities are unregulated, and also that entry to the fishery is uncontrolled. (These extreme assumptions are relaxed later.) The discussion follows McKelvey (1985); see also Clark et al. (2005).

First, net fleet operating revenue flow $R(t)$ is as before (Eq. 2.31):

$$R(t) = (pqx(t) - c)E(t) \tag{3.4}$$

Total fleet discounted operating revenue is

$$PV_0(E_{\max}) = \int_0^\infty R(t)e^{-\delta t}dt \tag{3.5}$$

where the subscript "0" indicates that the present value is assessed at time $t = 0$.

From the assumption that fishing effort is unregulated, we have (Eq. 2.35), i.e.

$$E(t) = \begin{cases} E_{\max} & \text{if } x(t) > x_{\text{OA}} \\ G(x_{\text{OA}})/qx_{\text{OA}} & \text{if } x(t) = x_{\text{OA}} \\ 0 & \text{if } x(t) < x_{\text{OA}} \end{cases} \tag{3.6}$$

where x_{OA} is the bionomic equilibrium biomass

$$x_{\text{OA}} = c/pq \tag{3.7}$$

We henceforth refer to $x_{\text{OA}} = c/pq$ as the variable-cost (or operating-cost) bionomic equilibrium.

Once having purchased his vessel, the owner's decision whether to fish depends only on his anticipated operating revenue $R(t)$. Equation (3.6) says that the owner will go fishing if and only if his operating revenue is positive. Of course, in practice there may be additional options, such as deciding to fish somewhere else, or to offer whale-watching cruises, etc. We ignore such possibilities here. (As an aside here, it is possible that $E_{\max} < G(x_{\text{OA}})/qx_{\text{OA}}$ i.e., the maximum effort capacity is too low to

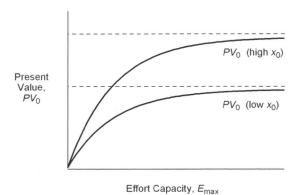

Figure 3.1 Present values of operating net revenues, PV_0, as a function of effort capacity, for two initial stock levels x_0.

reduce the biomass to variable-cost bionomic equilibrium. In this case we will have $E(t) = E_{\max}$ for all $t \geq 0$. As seen later, this case can actually occur in practice, if fixed costs are large.)

Thus, if $E_{\max}$ is known, we can calculate total discounted operating revenue PV_0 from Eq. (3.5). Figure 3.1 shows $PV_0 = PV_0(E_{\max})$, for two values of the initial stock level x_0. These curves tend asymptotically to a limit that corresponds to infinitely rapid fishing of the stock from x_0 down to x_{OA} (i.e. $E_{\max} \to +\infty$). The asymptotic limit is equal to

$$\int_{x_{\text{OA}}}^{x_0} (p - c/qx)dx = p(x_0 - x_{\text{OA}}) - \frac{c}{q}\ln\frac{x_0}{x_{\text{OA}}} \qquad (3.8)$$

and this is clearly an increasing function of the initial biomass x_0; higher initial biomass implies larger revenues from fishing down the stock.

The next question is how many vessels ($E_{\max}$) will enter the fishery? We begin by making a certain assumption about what happens to a fishing vessel over time. Namely, we assume that once purchased, vessels are permanently committed to the fishery in question. The vessels have no profit-producing alternative uses. Economists use the term "non-malleable capital" to refer to this situation.

On what basis will a potential vessel owner decide to enter the fishery? Let c_f (for fixed cost) denote the purchase cost of a vessel and let P denote the present value of future operating revenue that the new entrant expects to earn. Then entering the fishery will be advantageous if and only if $P > c_f$.

Now $P = PV_0(E_{\max})/E_{\max}$ where $E_{\max}$ is the existing fleet capacity, i.e., the number of vessels already in the fishery. Additional vessels will

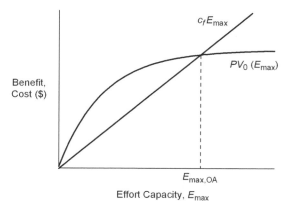

Figure 3.2 Investment equilibrium $E_{\text{max,OA}}$ in the open-access fishery.

therefore tend to enter the fishery if $PV(E_{\text{max}})/E_{\text{max}} > c_f$, and not otherwise. The number of entering vessels $E_{\text{max,OA}}$ will therefore satisfy

$$PV_0(E_{\text{max,OA}}) = c_f E_{\text{max,OA}} \qquad (3.9)$$

—see Fig. 3.2. The significance of this assumption will be discussed later. Equation (3.9) is an extension of the original idea of Gordon (1954), that fishermen will enter an open-access fishery if it is in their economic interest to do so. Gordon did not separate fixed and variable costs, however (see Smith 1969 for an alternative way of modeling entry and exit dynamics in open-access fisheries). By making this distinction explicit, our present model will allow us to consider the question of overcapacity and optimal capacity of fishing fleets.

Next we consider the possibility that no vessels will enter the fishery. This will be the case if $c_f E_{\text{max}} > PV_0(E_{\text{max}})$ for all $E_{\text{max}} > 0$. Recall that $PV_0(E_{\text{max}})$ depends on the initial biomass level x_0. If x_0 is too small, the fishery is too unprofitable to attract any new entrants.

To be specific, we show in Box 3.1 that entry will occur if and only if $x_0 \geq x_{\text{I}}$, where

$$x_{\text{I}} = \frac{c + \delta c_f}{pq} \qquad (3.10)$$

Thus x_{I} is an "investment threshold" stock level, in the sense that investment in fleet capacity will only occur if the current stock level is greater than or equal to x_{I}.

Box 3.1. Initial investment in the open-access fishery.

To begin with, suppose $x_0 = x_I$ as given by Eq. (3.10). Then if $E_{max} = E_I = G(x_I)/qx_I$ the fish population will stay in equilibrium at x_I and we have

$$PV_0(E_I) = \int_0^\infty (pqx_I - c)E_I e^{-\delta t} dt = \frac{(pqx_I - c)}{\delta} E_I = c_f E_I$$

by Eq. (3.10). Therefore by Eq. (3.9) we have $E_{max,OA} = E_I$ when $x_0 = x_I$. Also, we know that $E_{max,OA}$ is an increasing function of x_0, so that the curve S rises to the right of x_I as shown in Fig. 3.3.

To show that no investment occurs if $x < x_I$, note first that the first year's net operating revenue per vessel is at most $pqx_0 - c$, and we have $pqx_0 - c < pqx_I - c = \delta c_f$ by Eq. (3.10). Thus the would-be investor would not recover the interest on his investment during the first year of fishing. This explains why investment does not occur when $x_0 < x_I$. As shown in Fig. 3.3, investment thus only takes place for $x_0 \geq x_I$, up to the level given by the curve S.

Notice that x_I is analogous to $x_{OA} = c/pq$, which itself is a "fishing threshold" stock level, in the sense that fishing will not occur if the current stock level is below x_{OA}. We see that x_{OA} involves the variable cost of effort (c), whereas x_I involves the sum of variable cost plus the annual interest on fixed costs, δc_f.

To summarize, we have:

Table 3.1. *Fishing and investment under open-access.*

Current stock level x	Fishing occurs?	Investment occurs?
below x_{OA}	No	No
between x_{OA} and x_I	Yes	No
above or equal to x_I	Yes	Yes

There remains the question of how much investment in vessels will occur in the open-access fishery, given that the initial stock level x_0 is

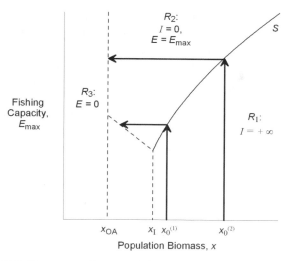

Figure 3.3 State-space diagram of the open-access fishery. The trajectories marked with arrows depict the development of the fishery starting from an initial biomass x_0. Two examples are depicted. (Vessel depreciation is here assumed to be zero; cf. Fig. 3.4.)

indeed larger than x_I. The answer $E_{\text{max,OA}}$ is an increasing function of the initial stock level x_0.

The resulting dynamics of the open-access fishery are shown in Fig. 3.3, which is a state-space diagram (McKelvey 1985). The two state variables are $x(t) =$ population biomass, and $E_{\text{max}}(t) =$ fleet capacity. The diagram also shows two typical (x, E_{max}) trajectories, as follows. Starting with initial biomass x_0 with $x_0 > x_I$, and assuming no previous investment, an initial investment increases capacity to the level $E_{\text{max,OA}}$ as specified by Eq. (3.9). As shown by Fig. 3.3, the larger x_0 is, the greater will be the initial capacity level $E_{\text{max,OA}}$; this is indicated by the curve labeled S in Fig. 3.3. Large, valuable, unexploited fish stocks tend to attract large initial investments of fishing capacity in an open-access fishery.

Following the initial investment, fishing activity proceeds according to Eq. (3.6):

$$E(t) = \begin{cases} E_{\text{max,OA}} & \text{if } x(t) > x_{\text{OA}} \\ E_{\text{OA}} & \text{if } x(t) = x_{\text{OA}} \end{cases} \tag{3.11}$$

where $x_{\text{OA}} = c/pq$ as before, and

$$E_{\text{OA}} = G(x_{\text{OA}})/qx_{\text{OA}} \tag{3.12}$$

Recall that x_{OA} is the variable-cost bionomic equilibrium, and E_{OA} is the corresponding level of effort that maintains this equilibrium. The lower trajectory drawn Fig. 3.3 does not actually reduce the stock level down to x_{OA}, but instead $x(t)$ ends up at a higher stock level $\bar{x}$ determined by setting $dx/dt = 0$, i.e., $G(\bar{x}) - qE_{\max,OA}\bar{x} = 0$. (The downward sloping dashed line in Fig. 3.3 is the line of these equilibrium solutions.) In this case, fleet capacity is not large enough to actually reduce the stock level to variable-cost bionomic equilibrium at x_{OA}. For a larger initial biomass x_0 (upper trajectory) the investment in fishing capacity is large enough to reduce $x(t)$ to x_{OA}, however.

To repeat, under the assumption of perfectly non-malleable fleet capital and instantaneous initial investment, our model of the unregulated open-access fishery predicts a sudden pulse of investment at time $t = 0$, followed by a phase of stock reduction, down to the level x_{OA}, where $x_{OA} = c/pq$ is the variable-cost bionomic equilibrium of the open-access fishery. No further investment occurs after the initial pulse. Fishing proceeds at full capacity until the stock is reduced to x_{OA}, at which time excess capacity ($E_{\max,OA} - E_{OA}$) exists. This excess capacity is not actually used on an annual basis, since vessels cease fishing whenever $x(t)$ falls below x_{OA}. (In the case exemplified by the trajectory starting at $x_0^{(1)}$ no excess capacity occurs at equilibrium.)

The State-Space Control Diagram

State-space control diagrams similar to Fig. 3.3 will appear elsewhere in this chapter, so it is important to understand what information they convey. First, note that the values of both control variables E (fishing effort) and I (investment rate) are specified in the figure, at every point $(x, E_{\max})$ of the state space. In Fig. 3.3 there are in fact three control regions R_i:

R_1 : region to the right of x_I and below the curve S
R_2 : region to the right of x_{OA} and left of x_I or above S
R_3 : region to the left of x_{OA}

The predicted (open-access) values of E and I are:

$$R_1 : I = +\infty, \ E \text{ not specified}$$
$$R_2 : I = 0, E = E_{\max} \qquad\qquad (3.13)$$
$$R_3 : I = 0, E = 0$$

The example trajectories drawn in Fig. 3.3 are deduced from these control rules. Other trajectories, initiated at any chosen point, can be deduced in the same way.

To complete the control diagram, we need to specify what controls are used on the lines $x = x_{OA}$ and $x = x_I$:

$$x = x_{OA} : I = 0$$
$$E = \begin{cases} E_{OA} & \text{if } E_{max} \geq E_{OA} \\ E_{max} & \text{if } E_{max} < E_{OA} \end{cases} \qquad (3.14)$$

$$x = x_I, E_{max} < E_I : I = +\infty \text{ (increasing } E_{max} \text{ to } E_I) \qquad (3.15)$$

Here E_{OA} is given by Eq. (3.12) and $E_I = G(x_I)/qx_I$. These special cases are also derived in Box 3.1.

As an illustration, let us deduce the trajectory starting from an initial point (x, E_{max}) lying between x_{OA} and x_I, and below E_I. Here we have $I = 0$ and $E = E_{max}$, so the trajectory moves horizontally to the right $(dx/dt > 0)$ until it hits the vertical line $x = x_I$. At this point there is a pulse investment $(I = +\infty)$ that increases E_{max} up to the level E_I. Thereafter the system remains in equilibrium at the point (x_I, E_I). This example would apply to the case of a previously heavily fished stock with a small existing fishing fleet, such as might have occurred just after the establishment of a 200-mile zone. Domestic fishing would only develop (unless subsidized) after the stock had recovered to x_I, at which time E_I vessels would enter the fishery. (Alternatively, the government might opt to lease fishing rights to foreign vessel owners, who would be willing to enter the fishery immediately because their capital costs have already been paid.)

The control diagram in Fig. 3.3 is admittedly a bit complicated. Also, it depends on several questionable model assumptions (see below). Nevertheless the main predictions seem to agree qualitatively with commonly noted characteristics of unregulated open-access fisheries, including:

- Rapid expansion of fishing capacity will take place in a newly developing fishery.
- Early reduction of the standing stock of fish will then occur.
- Ultimate equilibrium will occur, with a reduced fish stock and with low or zero annual net operating revenues at equilibrium.
- Excess fishing capacity (i.e. capacity larger than necessary for the annual catch) will exist at this equilibrium (except for the case where $\bar{x} > x_{OA}$).

The following sensitivity results also seem reasonable:

- An increase in capital costs c_f will cause x_I to shift right, and the curve S to shift right and down. Thus capacity $E_{\max,OA}$ will be smaller.
- An increase in the discount rate δ will have the same effect as an increase in c_f.
- An increase in the price p of fish will cause both x_{OA} and x_I to shift left, and S to shift left and up. Thus capacity $E_{\max,OA}$ will be larger, and the fish stock will be more heavily exploited.
- An increase in the population's intrinsic growth rate r will cause the equilibrium line in Fig. 3.3 to shift up, but will not affect x_{OA} or x_I. (The equilibrium line has equation $rx(1 - x/K) - qE_{\max}x = 0$, or $E_{\max} = (r/q)(1 - x/K)$.) The curve S will also shift up, i.e., fishing capacity $E_{\max,OA}$ will be larger.

Perfect Foresight

Our basic equation for predicting initial fleet capacity $E_{\max,OA}$ is Eq. (3.9). This equation is based on two simplifying assumptions, both of which may be unrealistic. First, investment in fleet capacity is assumed to occur instantaneously at $t = 0$. In actuality, the build-up of capacity is likely to be somewhat gradual. However, a strong incentive exists for fishermen to enter a newly established fishery quickly, before other vessel owners have garnered the large initial profits from the fishery. Thus in fact a rapid expansion of capacity may occur and indeed often does occur in practice. But even if the initial investment is gradual, Fig. 3.3 still applies, except that the trajectories now bend to the left as they rise towards the curve S. A gradually expanding fleet begins to fish down the stock, and investment ceases when the trajectory hits S.

The second assumption is based on perfect foresight, meaning that each potential vessel owner correctly forecasts the entire future of the fishery in question. Namely, the owner correctly calculates his future costs and revenues and also correctly predicts how many other vessels will actually enter the fishery. Furthermore, all potential vessel owners perform this calculation instantaneously. This assumption is obviously quite unrealistic.

In actuality, possible owners may over- or underestimate the economic potential of the fishery, resulting in initial investment either higher or lower than predicted by the perfect foresight model. For example, if initial capacity $E_{\max}$ is too large, vessel owners will lose money on their investments, but at least this will tend to discourage further expansion.

In spite of such reservations, the qualitative predictions described earlier should remain valid.

Before proceeding let us briefly review and critique our open-access investment model. We imagine that a novel fishing opportunity has just arisen, either from the discovery of a previously unexploited resource, or because of new fishing technology, or because of new market opportunities. Potential fishermen have to decide whether to purchase a suitable vessel and enter the fishery. In making this decision a fisherman will compare the cost of his investment with his anticipated revenues from the fishery. These revenues will depend on price and cost parameters, but also on how many other vessels will enter the fishery, and how long the fish stock will remain at a profitable biomass level. These are exactly the decision components that our mathematical model attempts to address. Even if the potential fisherman does not construct a mathematical model, he will nevertheless try to forecast future developments in the fishery.

Of course there will be many uncertainties in this calculation, not the least of which may be the probability of future government regulation. The latter question will be taken up later in this chapter and in Chapter 4.

Vessel Depreciation

Another important assumption in the above model is that vessels, once acquired, never deteriorate or depreciate. This is also unrealistic. If we instead assume a constant annual rate of depreciation γ, Fig. 3.3 changes as shown in Fig. 3.4. A brief description follows.

First, the investment threshold x_I now becomes

$$x_I = \frac{c + (\delta + \gamma)c_f}{pq} \tag{3.16}$$

This is the same as Eq. (3.10), with an additional cost γc_f, the annual cost of vessel depreciation. As before, investment in fleet capacity occurs provided $x_0 \geq x_I$, but not otherwise. Once the initial investment has been completed, the fishery operates at full capacity. However, fleet capacity gradually decreases (depreciates), eventually to such a low level that $dx/dt > 0$, i.e., the fish stock begins to recover. When $x(t)$ hits x_I on the recovery path, investment in additional capacity again occurs, and the fishery reaches a long-run equilibrium at the position (x_I, E_I). Here, annual depreciation is made up with continuous re-investment at rate γE_I.

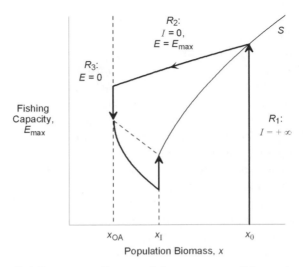

Figure 3.4 State-space diagram of the open-access fishery, assuming a positive rate of vessel depreciation

The revised model thus predicts a dynamic cycle of fleet capacity $E_{\max}(t)$ and fish population biomass $x(t)$, which ultimately reaches a long-term equilibrium $(x_{\mathrm{I}}, E_{\mathrm{I}})$. We refer to x_{I} as the total-cost bionomic equilibrium biomass, whereas x_{OA} is the variable-cost bionomic equilibrium. Thus x_{OA} involves only variable effort costs, while x_{I} involves these variable costs plus interest and depreciation on capital.

A similar, counterclockwise cycle of fleet capacity and stock biomass was predicted by Smith (1969), based on a model assuming that the rate of entry or exit of vessels is proportional to current operating revenues $pqx - c_{\mathrm{total}}$ where c_{total} denotes total effort costs. Thus fleet size increases when $x > x_{\mathrm{I}}$, and decreases when $x < x_{\mathrm{I}}$. Berck and Perloff (1984) obtain a similar prediction assuming rational future expectations. Neither of these models considers fixed costs of vessels per se. Smith's prediction has been successfully tested, qualitatively, using data from various fisheries (McKelvey 1986). These tests cannot distinguish, however, between the Smith model and the McKelvey model used here. The latter model is convenient for studying such questions as overcapacity and its reduction, as we shall see.

The existence of an investment threshold x_{I} (with or without depreciation) implies a degree of investment stability in the event of changes in fish price or cost of fishing. For example, suppose that the price p increases unexpectedly at some time $t_1 > 0$. This price change implies new, lower threshold values x'_{OA} and x'_I. To predict whether this will induce an expansion of fleet capacity E_{max}, we need to redraw our state-space diagram (Fig. 3.3 or 3.4), and determine which modified region R'_i the current state point $(x(t_1), E_{max}(t_1))$ lies in. It is easy to see that, even if for example $x(t_1) > x'_I$ new investment may not occur. The reason, of course, is that the current fish stock $x(t_1)$ is not large enough to attract investment over and above the current fleet size $E_{max}(t_1)$. In this case the original entrants will enjoy greater benefits than anticipated.

The original entrants, however, might have expected that the price of fish would rise over time. If so, the initial entry E_{max} would be larger than predicted by our model, resulting in further rent dissipation. Later in this and the next chapter we will show how anticipated changes in fishery management may often encourage further over-expansion of capacity.

Non-malleability Assumption

Another important assumption of the above models is that the initial fishing fleet, of size E_{max}, is never reduced in size, except through depreciation. All vessels that enter the fishery are "trapped" there, and never transfer to other fisheries, or leave to be used for some other purpose (or to be scrapped). Except for gradual reduction of capacity through depreciation, vessel capital is assumed to be non-malleable in this sense.

Two alternative assumptions to complete non-malleability are, first, partial malleability, and second, full malleability. Under full malleability (i.e. fully reversible investment) the initial cost of a vessel can be fully recovered (less depreciation) whenever the vessel leaves the fishery. In these circumstances there are no fixed costs whatever, since vessels can be "rented" at any time. The rental cost rate is equal to $(\gamma + \delta)c_f$, i.e. the annual costs of depreciation and interest. Hence there is now just a single bionomic equilibrium, at $x = x_I$ as given by Eq. (3.13). Also, there is now no problem of permanent overcapacity, since vessels will automatically be withdrawn whenever $x(t) < x_I$.

What about initial entry $E_{max,OA}$, i.e., the position of the curve S in Figs. 3.3 and 3.4, under full malleability? What happens is that S now becomes a vertical line—the model predicts unlimited initial capacity. This follows from applying Eq. (3.9) with $c_f = 0$; fixed costs of investment are zero if vessel capital is perfectly malleable.

In other words, with perfectly malleable vessel capital, our model predicts that a newly discovered, valuable fish stock will experience a large-scale, short-lived fishery by a fleet of mobile vessels. The likelihood is all the greater if such a fleet is already in existence, perhaps as the result of previous fishing activities. An example of this process was the rapid, sequential depletion in the 1990s of some stocks of orange roughy, a commercially valuable deep-water species that forms dense spawning aggregations (Munro et al. 2004). Being long-lived (up to 150 years, with sexual maturity at around 30 years), orange roughy are subject to severe depletion and slow recovery. Several stocks of orange roughy off Australia, New Zealand and South Africa were fished out over short periods in the 1980s and '90s.

Another instance of mobile, or malleable fleet capacity occurred in the history of distant-water factory fleets in the years preceding the declaration of EEZs (Extended Economic Zones) in the late 1970s. Large fleets of trawlers and factory vessels moved from one off-shore fishery to another, in a process referred to as pulse fishing. These fleets severely depleted important stocks of hake, cod and other species. Much of the motivation behind the establishment of EEZs may have come from these developments.

Partial Malleability

Under partial malleability vessels can be withdrawn from the fishery, but at a re-sale price $c_w < c_f$. For example, a vessel may switch to, or be sold for use in another fishery, requiring extensive refitting. Or the vessel may simply be sold as scrap metal. Under this assumption a new disinvestment threshold $x_D = (c + (\delta + \gamma)c_w)/pq$ occurs in Fig. 2.4. We omit the details (see McKelvey 1985).

Which is the more realistic assumption—malleability or non-malleability—depends to a large extent on the degree of specialization of the vessels involved, and may also depend on the availability of alternative fish stocks. It seems unlikely, however, that most vessel purchasers take into account the possibility of selling or switching later. Foresight can only go so far. Whatever the reason, non-malleable overcapacity is doubtlessly widespread in today's fishing industry.

An Alternative Catch Equation

We argued in Sec. 2.1 that the Schaefer catch equation $h = qEx$ is likely to be misleading for many fish species. For the alternative equation $h = qE$ it was shown that if the fishery is economically viable then bionomic equilibrium occurs at $x = 0$, i.e. biological extinction. Let us briefly review and extend this argument.

Net fleet operating revenue is now given by

$$R(t) = (pq - c)E(t) \quad \text{if } x(t) > 0 \qquad (3.17)$$

and this remains positive independent of stock level x, provided that

$$pq - c > 0 \qquad (3.18)$$

This condition is both necessary and sufficient for a viable fishery, and when it holds, variable-cost bionomic equilibrium occurs at $x = 0$.

But what about fleet capacity? Could the cost of a fishing vessel be so high as to discourage entry into the fishery? Our full model (see Eqs. (3.2)–(3.5)) is now

$$\frac{dx}{dt} = G(x) - qE \quad (x > 0) \qquad (3.19)$$

$$0 \leq E(t) \leq E_{\max} \qquad (3.20)$$

$$PV_0(E_{\max}) = \int_0^\infty R(t)e^{-\delta t}\,dt \qquad (3.21)$$

(where the upper limit of integration is replaced by $T = $ time that stock $x(t)$ reaches extinction, if this does occur). We now assume that Eq. (3.18) holds. We also assume that

$$E(t) = \begin{cases} E_{\max} & \text{if } x(t) > 0 \\ 0 & \text{if } x(t) = 0 \end{cases} \qquad (3.22)$$

There are two cases to consider, depending on whether $qE_{\max} < h_{\max} = \max_x G(x)$, or the reverse. In the first case we see that $x(t) \to \bar{x} > 0$ as $t \to \infty$, so that $R(t) = (pq - c)E_{\max}$ for all $t > 0$. Hence

$$PV_0(E_{\max}) = \int_0^\infty R(t)e^{-\delta t}\,dt$$

$$= \frac{(pq - c)E_{\max}}{\delta} \quad \text{if } qE_{\max} < h_{\max} \qquad (3.23)$$

In case two, we have $x(t) \to 0$ in finite time (depending on the initial stock level x_0). This implies that $PV_0(E_{\max})$ is less than the expression given in Eq. (3.23). The situation is depicted in Figure 3.5. As before, under perfect foresight we predict that fleet capacity under open access will reach the level $E_{\max,\mathrm{OA}}$ determined by

$$PV_0(E_{\max,\mathrm{OA}}) = c_f E_{\max,\mathrm{OA}} \qquad (3.24)$$

as shown in Fig. 3.5. Note that

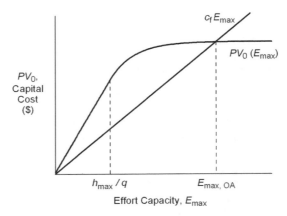

Figure 3.5 Present value $PV_0(E_{\max})$ and capital cost $c_f E_{\max}$ as functions of fleet capacity $E_{\max}$, for the model using the alternative catch equation.

$$E_{\max,\mathrm{OA}} \begin{cases} = 0 & \text{if } pq - c < \delta c_f \\ > h_{\max}/q & \text{if } pq - c > \delta c_f \end{cases} \qquad (3.25)$$

(This follows by noting that $PV_0(E_{\max})$ is linear for $E_{\max} < h_{\max}/q$, with slope $(pq - c)/\delta$; see Eq. 3.23.) We therefore reach the following conclusion:

Under the alternative catch relationship $h = qE$, the fishery is economically viable if and only if

$$pq > c + \delta c_f \qquad (3.26)$$

If this is the case the open-access fishery will drive the stock to extinction in finite time.

This is the same prediction derived in Chapter 2, but now both variable costs (c) and amortized fixed costs (δc_f) are explicitly represented in the model.

A Less Extreme Model

As noted in Chapter 2, the alternative model $h = qE$ is an extreme case, based on the assumption of complete density-independence of catch rates. Most actual fisheries probably are best described by a Type I CPUE profile, for which the power law

$$h = qx^\beta E \quad (\beta < 1) \qquad (3.27)$$

is a first approximation (Cooke and Beddington 1985). Recall that this equation includes $h = qxE$ and $h = qE$ as limiting cases, for $\beta = 1$ and $\beta = 0$ respectively. Let us briefly review the above calculations, Eqs. (3.2)-(3.13), for this case; this is quite straightforward.

First, we now have

$$R(t) = (pqx^\beta - c)E(t) \qquad (3.28)$$

Hence variable-cost bionomic equilibrium is now

$$x_{\text{OA}} = \left(\frac{c}{pq}\right)^{1/\beta} \qquad (3.29)$$

Thus extinction is ruled out in this model (if $\beta > 0$), for the same reason as in the case of the Schaefer model—namely, the catch rate approaches zero as $x \to 0$, so fishing becomes uneconomical below x_{OA}, which is positive.

The investment threshold is now given by

$$x_{\text{I}} = \left(\frac{c + (\gamma + \delta))c_f}{pq}\right)^{1/\beta} \qquad (3.30)$$

With these changes, Fig. 3.4 again applies. Thus the theory of non-malleable investment is qualitatively the same as in the Gordon–Schaefer model ($\beta = 1$). Quantitatively, the threshold levels x_{OA} and x_{I} decrease as $\beta \to 0$. Also, since fishing remains profitable down to low stock levels, the predicted fleet capacity $E_{\text{max,OA}}$ increases as $\beta \to 0$. Thus a strongly Type I fishery may motivate greater fleet capacity than otherwise, increasing the likelihood of resource depletion.

Consequences of Anticipation

Anticipation of future revenues is the essence of investment decisions. In fisheries, a potential vessel owner attempts to assess his future operating income before deciding whether to obtain a vessel. In the open-access situation vessels will continue to enter the fishery as long as estimated discounted benefits exceed the cost of a vessel. Because the resource is finite, benefits per vessel decrease with the number of vessels, and it is this feature that limits total entry to the fishery. Also, the larger the initial resource stock the more vessels will enter the fishery.

Total benefits also depend on how the fishery will be managed. Until now we have assumed that no management whatever will occur. In reality, most commercial fisheries are now managed, usually with biologically based objectives such as MSY. What effect will such anticipated

management have on fishermen's investment decisions? For the moment
we continue to assume that entry to the managed fishery remains unrestricted; Homans and Wilen (1997) refer to this situation as "regulated
open access."

The effect of fishery management on fleet capacity will depend on
whether the management will increase or decrease future revenues. Ideally one would hope that by preventing overfishing, management would
enhance fishermen's revenues, but as demonstrated in Chapter 2 this
may not be the case because of discounting.

Let us now assume, however, that management will in fact increase
the discounted present value of future operating revenues. If access to
the fishery remains unregulated, the result will be an increase in fleet size
relative to the case of an unmanaged fishery. This increase in capacity
will in fact dissipate all of the potential economic benefits from future
management. I propose that this is a fundamental principle of regulated
open-access fisheries:

> *Under regulated open access, all potential economic benefits of*
> *management are dissipated through excess expansion of fishing capacity.*

The implications of this principle will permeate much of the remainder
of this book.

Limited Entry

The solution to the problem of overcapacity in fisheries, regulated or
otherwise, would seem to be to limit entry to the fishery. Doing so has
often proved more difficult than expected, however. For example, James
(2004) reports that three consecutive attempts to reduce and control
fleet size in the British Columbia salmon fishery had almost no effect
in terms of economic benefits. We take up the question of controlling
fishing capacity in Sec. 3.4.

3.3 A Dynamic Optimization Model

We next develop an optimization model that includes harvesting and
investment strategies. The following model equations are closely parallel
to our previous model of Eqs. (3.2)-(3.5).

$$\frac{dx}{dt} = G(x) - qEx, \quad x(0) = x_0 \tag{3.31}$$

$$0 \leq E(t) \leq K(t) \tag{3.32}$$

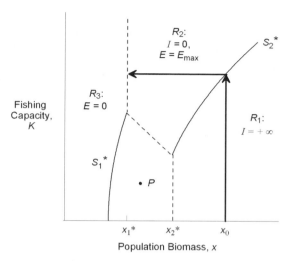

Figure 3.6 Feedback control diagram for the dynamic investment optimization model. See text and Box 3.2.

$$\frac{dK}{dt} = I, \quad K(0) = K_0 \tag{3.33}$$

$$0 \leq I(t) \leq +\infty \tag{3.34}$$

$$R(t) = (pqx(t) - c)E(t) \tag{3.35}$$

$$PV_{\text{opt}} = \max_{E(t), I(t)} \int_0^\infty (R(t) - c_f I(t))e^{-\delta t} dt \tag{3.36}$$

Equations (3.31) and (3.32) are the same as Eqs. (3.2) and (3.3) with a change of notation (for simplicity) $E_{\max} = K$. We think of $K(t)$ as the capital invested in fleet fishing capacity. The investment rate $I = I(t)$ is now a control variable, to be chosen in an optimal manner, to maximize net present value, as given by Eq. (3.36).

The foregoing is a two-state and two-control-variable optimal control problem. In general, such problems can be notoriously difficult to solve in explicit form. In the present case, the solution is at least fairly easy to describe. A mathematically rigorous proof of the correctness of this solution appears in Clark et al. (1979).

The optimal solution is indicated in Fig. 3.6. Box 3.2 describes the solution in detail. Here we give a brief description. Note first that there are again two distinguished biomass levels, denoted by x_1^* and x_2^*. The first value x_1^* is the optimal variable-cost biomass; in fact x_1^* is the same

Box 3.2. Solution of the dynamic optimal investment model (zero depreciation).

Consider the optimization problem of Eq. (3.36). First, there are two distinguished biomass levels x_1^* and x_2^*, determined by the equations

$$G'(x_i^*) - \frac{c_i'(x_i^*)G(x_i^*)}{p - c_i(x_i^*)} = \delta \quad (i = 1, 2) \tag{3.37}$$

where

$$c_i(x) = \frac{c_i}{qx} \tag{3.38}$$

Here we have

$$c_1 = c \text{ and } c_2 = c + \delta c_f \tag{3.39}$$

(compare Eq. 2.63). Investment takes place only for $x_0 \geq x_2^*$. The optimal level of investment is specified by the curve labeled S_2^* in Fig. 3.6. Once this initial investment K has been made, fishing proceeds at full capacity $(E(t) = K)$ as long as $x(t) > x_1^*$. Specifically

$$E(t) = \begin{cases} K & \text{if } x(t) > x_1^* \\ G(x_1^*)/qx_1^* & \text{if } x(t) = x_1^* \end{cases}$$

Thus the biomass is gradually reduced, either to the target equilibrium x_1^*, or if capacity K is not large enough, to an intermediate level $\bar{x}$.

Figure 3.6 is a full feedback control diagram, which specifies the optimal control values $E(t)$ and $I(t)$ for any given current state combination $x(t), K(t)$. For example, if the state point (x, K) lies in Region R_3 to the left of the curve labeled S_1^* (or left of $x = x_1^*$), then $E(t) = 0$: the fishery should be closed. In this situation the stock has been severely depleted, and should be rehabilitated as rapidly as possible by a moratorium. (Why the switching curve S_1^* actually lies to the left of x_1^* is a technical point related to the theory of irreversible investment, as described by Arrow (1968) in a different setting. See Clark et al. 1979.) If the system state (x, K) lies in R_2 between the two curves S_1^* and S_2^*, the optimal effort is $E(t) = K$ (and of course $I(t) = 0$). For example, imagine that a coastal nation takes over a fishery that has been severely depleted by a fleet consisting of domestic and foreign vessels, the latter being then expelled from the fishery. Suppose the state of the system

---

Box 3.2 *continued*

is now at point P in Fig. 3.6, with a small domestic fleet, and a depleted stock. The optimal strategy is then to utilize this fleet fully, which will allow the stock to recover gradually towards the level x_2^*.

When the stock reaches x_2^* it then becomes optimal to invest in additional vessels, building the fleet up to $K_2^* = G(x_2^*)/qx_2^*$. Thereafter the fishery remains in a steady equilibrium state, with $x = x_2^*$ and $K = K_2^*$. Finally if (x, t) lies in region R_1, right of (or at) x_1^* and below S_2^*, the optimal strategy is an instantaneous investment that increases capacity to K^* as specified by the curve S_2^*. See Clark et al. (1979) for full details.

---

as x_δ in Eq. (2.54). The second value x_2^* is the optimal target biomass when the interest δc_f on capital is included in the cost of effort (see Eqs. (3.37)–(3.39)).

The optimal level of initial investment depends on the initial stock level x_0, as indicated by the curve S_2^* in Fig. 3.6. In particular, no investment is made unless $x_0 \geq x_2^*$. Once the initial investment K has been made, the fishery proceeds with $E(t) = K$. The fish stock is then gradually fished down to a long-term equilibrium $\bar{x}$, where

$$x_1^* \leq \bar{x} \leq x_2^*$$

Note that the discount rate δ actually plays a double role in the present model: it is the rate at which future revenues are discounted, and also the rate of interest earned on capital (the so-called opportunity cost of capital). Perhaps these should be treated as two different rates, although this might appear to be a contradiction. In any event, the conclusion does raise questions about the general validity of the rule that high discount rates always imply depletion of renewable resource stocks. Certainly we can no longer equate the outcome of infinite discounting with open-access conditions, if costs of capital are taken into consideration. Indeed, the model predicts that no fishing is optimal if the discount rate δ is too large.

Although the control diagram for the optimization model looks rather similar (graphically speaking) to the open-access case (Fig. 3.3), the biomass levels x_i^* are now determined by an optimality condition, rather than being instances of bionomic equilibrium. By comparing Eq. (3.37) in Box 3.2 with Eqs. (3.7) and (3.10), we see that

$$x_1^* > x_{OA} \quad \text{and} \quad x_2^* > x_I \tag{3.40}$$

In other words, the optimal variable-cost biomass level x_1^* is higher than the bionomic equilibrium, and the optimal investment threshold x_2^* is higher than the open-access investment threshold. Optimal fishing is less damaging to the fish population than open-access fishing. This is hardly surprising! In addition, of course, optimal fishing and investment preserves—indeed maximizes—economic yield from the fishery, yield that would be dissipated under open-access.

Vessel Depreciation
The above model assumes zero depreciation of vessel capital K. To include depreciation, we modify Eq. (3.33) as follows

$$\frac{dK}{dt} = I - \gamma K \tag{3.41}$$

where γ denotes the depreciation rate. This is the same modification that we made in the open-access model. Figure 3.7 shows the effect of this modification.

Again there is an optimal initial investment, with $K(0)$ specified by the curve S_2^*. Thereafter $E(t) = K(t)$, as before, but now the capacity $K(t)$ slowly declines through depreciation. The fishery goes through a cycle of high initial capacity, resulting in rapid stock depletion, but followed by a phase of declining capacity and eventual stock recovery. Finally a long-term equilibrium is reached at $x = x_2^*$ and $K = K_2^*$, following a further investment, as indicated. It should be emphasized that this cycle is economically optimal, according to the present model.

The value of x_2^*, which is both the threshold biomass for investment in vessels, and the long-term optimal equilibrium biomass, is again determined by Eqs. (3.37)–(3.38), but with cost function

$$c_2(x) = \frac{c + (\gamma + \delta)c_f}{qx} \tag{3.42}$$

Thus x_2^* takes account of both interest and depreciation of capital. At this long-term equilibrium we have $dK/dt = 0$, so that

$$I^* = \gamma K_2^* \text{ where } K_2^* = G(x_2^*)/qx_2^* \tag{3.43}$$

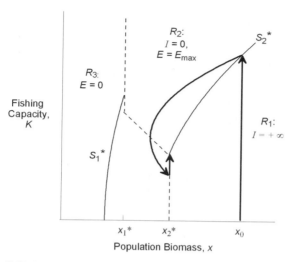

Figure 3.7 Feedback control diagram for the dynamic investment optimization model with depreciation. The heavy curve with the arrows is a typical optimal dynamic trajectory of fleet capacity $K(t)$ and population biomass $x(t)$ over time. The counter-clockwise cycle is characteristic of a renewable resource industry. From Clark et al. (1979).

At equilibrium, investment occurs at a constant rate, just matching depreciation of fleet capacity.

Clark and Lamberson (1982) applied the dynamic optimization model (with depreciation) to the postwar Antarctic whaling industry catching blue, fin, and sei whales. They calculated an optimal initial postwar (1946) investment of $K_0 = 9$ factory fleets, compared to the long-run optimal equilibrium of $K_2^* = 1$ factory fleet. The time to reach equilibrium at $x_2^* = 120,000$ BWU (blue-whale units) was 31 years. (Value of $\gamma = .15/\text{yr}$, $\delta = .10/\text{yr}$.) This compares with the actual postwar history of Antarctic whaling, in which fleet size expanded to 21 factory fleets, which reduced whale stocks from 260,000 to about 80,000 BWU over 15 years. This fleet size of 21 units is in fact considerably smaller than predicted by the open-access model, namely 34 factory fleets (McKelvey 1986). Fleet expansion was not instantaneous, but took place gradually over the postwar years 1946–1961. Doubtlessly the early entrants garnered large initial profits from fishing down the whale stocks, resulting in a reduced incentive for later entry of additional whaling fleets.

In any event, the whaling example serves to indicate the extent of overcapacity that can occur in practice. First, the optimal (postwar) initial capacity equals 900% of the long-term optimal equilibrium. The theoretical open-access initial capacity is 3400%, and the actual built capacity 2100% of the optimal equilibrium. This example may not be particularly unusual.

Management Implications

Our dynamic optimization model has several novel management implications:

1. Optimal fleet capacity does not necessarily coincide with minimal capacity, that is, the minimum capacity required to capture the optimal sustained yield. The large revenues obtainable during the initial stages of a new fishery can justify the acquisition of more vessels than will be needed later on. Unless vessel capital is completely malleable, this excess capacity may optimally remain active in the fishery. (Exceptions to this statement are discussed in Chapter 4.)

2. Nevertheless, optimal initial capacity is often much smaller than initial capacity under open-access. Thus a newly developing fishery should ideally be regulated to prevent undue overexpansion of fishing capacity. Unfortunately, effective regulation of a new fishery is seldom practical, largely for lack of information about the magnitude of the resource and about its biological properties. Excess capacity may therefore be all but inevitable in newly developing fisheries. Finding effective ways to reduce overcapacity is thus a high priority issue in fisheries management. We discuss this topic in detail in the next section.

3. Excess fleet capacity, if it exists (from whatever cause) should not necessarily be entirely eliminated, or prevented from participating in the fishery. However, annual catch quotas (TACs) should be specified to prevent overfishing, relative to the short-term equilibrium stock level (x_1^*).

Given that the above optimization model employs numerous simplifying assumptions, many of which may be unrealistic, what relevance do these model-derived prescriptions have for real-world fisheries? Let us focus on a typical situation: existing capacity in a certain fishery is far in excess (e.g., by an order of magnitude) of K^*, the level of capacity needed to capture the estimated optimal sustainable yield. The fishery is about to be "rationalized" by means of TACs. The present biomass level x is far below the long-term optimum. Should the fishery be (a) shut down for an extended period of stock recovery, or (b) operated at a low TAC, allowing a more gradual recovery of the fish population, or (c) allowed to operate at full effort capacity, as long as stocks do not decline below some predetermined critical level?

These are difficult choices. Shutting down the fishery may have devastating consequences for fishermen, who may have limited alternative employment opportunities (human capital, as well as vessel capital, may be non-malleable). Yet allowing the fishery to operate at full capacity may place the fish population at risk of collapse. To add to the difficulty, the over-expanded fishing industry typically presses for continuing large catch quotas, citing economic need. Managers may have little hard evidence that severe quota reductions are necessary to preserve the stock. The resulting decision may be a compromise that neither protects the resource nor satisfies the fishermen. No matter what happens the managers will be sharply criticized.

No amount of "science" is going to resolve this dilemma. Later chapters will discuss possible approaches that seem to hold some promise of overcoming current problems in fisheries management. To foreshadow this later material, let me mention two cornerstones of a strategy for improved management: individual fishing quotas, and adequately sized marine protected areas. Not long ago both these measures were generally considered to be outlandish fairy-tale ideas. The continuing failure of traditional management approaches, however, has increased the degree of interest in such unorthodox approaches. Individual quotas are already in place in many marine fisheries; protected areas have not been widely implemented as yet, although the concept is now attracting growing support.

Alternative Catch Equation
Suppose now that

$$h = qx^\beta E \quad \text{where } 0 < \beta < 1 \tag{3.44}$$

Then the unit cost function of Eq. (3.39) becomes

$$c_i(x) = \frac{c_i}{qx^\beta} \tag{3.45}$$

The rest of the optimization model is unchanged. The threshold stock levels x_i^* are again given by Eq. (3.38) and the state-space control diagram is similar to Fig. 3.7.

These results change, however, if $\beta = 0$, i.e., if the catch rate is independent of stock size. The harvesting cost functions are now

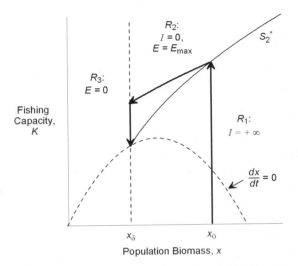

Figure 3.8 Feedback control diagram for the dynamic optimal investment model, using the alternative catch equation $h = qE$.

$$c_1(x) = c, \quad c_2(x) = c + (\delta + \gamma)c_f \qquad (3.46)$$

These are also independent of the stock size x. Hence Eq. (3.37) becomes

$$G'(x_i^*) = \delta \quad (i = 1, 2) \qquad (3.47)$$

Thus the two biomass levels x_i^* coincide, and are equal to x_δ as obtained in Chapter 2 for this case. The feedback control diagram changes, as shown in Fig. 3.8. The fishery is economically viable if and only if

$$pq > c + (\gamma + \delta)c_f \qquad (3.48)$$

If this is the case, the optimal strategy is to invest in vessels up to the level indicated by curve S_2^* in Fig. 3.8, and then to fish at full capacity until the biomass $x(t)$ reaches x_δ. Subsequently, equilibrium is maintained at $x = x_\delta$, with $E = E_\delta = F(x_\delta)/q$. There exists temporary excess capacity, until $K(t)$ declines through depreciation to the level $K_\delta = E_\delta$. In contrast to the previous model, here there is no non-malleability gap. Once the biomass is reduced to the long-term equilibrium at x_δ, excess capacity exists but is not utilized.

3.4 Regulated Open Access and Buy-Back Programs

The rest of this chapter is devoted to a currently popular method of reducing capacity, vessel buy-back programs. As noted earlier, governments have already spent hundreds of millions, if not billions, of dollars on buy-backs, with limited success. We will argue here that, unless done right, buy-back programs are pretty much doomed to failure, and may actually worsen the overcapacity problem in the long run. How to achieve capacity reduction without buy-backs will be investigated in Chapter 4.

Because buy-back programs are usually used in fisheries that are already being managed using TACs (total allowable catch quotas), we henceforth assume that this is the case. The annual catch quota is denoted by Q, here assumed to be a constant. We will develop a simple model to deal specifically with this situation. The current biomass level is x, and we assume (simplistically) that the TAC system succeeds in maintaining x as a constant stock level. To further simplify the model we will suppose that the annual catch H is given by

$$H = qEx \qquad (3.49)$$

where E denotes annual effort, measured in standardized vessel years. Equation (3.49) is an approximation that ignores annual changes in the stock level caused by fishing, natural mortality, and recruitment. These simplifying assumptions make the model tractable, without seriously affecting the conclusions. (The most damaging assumption here is probably the neglect of random natural fluctuations in stock abundance. We discuss this question in Chapter 5.)

The TAC constraint is

$$H \leq Q \qquad (3.50)$$

Let K again denote fleet capacity (SVU). Also, let D_1 denote the maximum possible length of the fishing season, as a fraction of the full year. The actual length of the fishing season, as specified by the managers, is D years, where

$$0 \leq D \leq D_1 \qquad (3.51)$$

Assuming that full fleet capacity is used during the fishing season, we have

$$E = KD \qquad (3.52)$$

$$H = qKDx \qquad (3.53)$$

If the quota Q is caught we have $H = Q$, and the season length D is

$$D = \frac{Q}{qKx} \quad \text{if } H = Q \qquad (3.54)$$

Thus a large fleet K forces the managers to set a short season D. The fleet size $\hat{K}$ that would be just sufficient to capture the annual quota Q in the full season D_1 is obtained by putting $D = D_1$ in Eq. (3.54):

$$\hat{K} = \frac{Q}{qD_1x} \qquad (3.55)$$

For the present discussion we will suppose that $\hat{K}$ is the optimal fleet size (keeping in mind that in practice a somewhat larger fleet may be optimal, for various reasons).

To help understand the model, we consider a numerical example, with

$$Q = 10,000 \text{ tonnes}$$
$$x = 50,000 \text{ tonnes}$$
$$q = .01/\text{SVU year}$$
$$D_1 = 0.5 \text{ years}$$

Then $\hat{K} = 40$ SVU, i.e., 40 vessels are sufficient to catch the annual quota.

Annual fleet operating revenue is

$$R = pH - cE = (pqx - c)KD \qquad (3.56)$$

Note that this is a constant for $K \geq \hat{K}$, for then we have $KD = Q/qx$. Hence

$$R = \bar{R} = (p - \frac{c}{qx})Q \quad \text{if } K \geq \hat{K} \qquad (3.57)$$

For $K < \hat{K}$ we have $D = D_1$ so by Eq. (3.56)

$$R = (pqx - c)KD_1 \quad \text{if } K < \hat{K} \qquad (3.58)$$

The present value of fleet operating revenue is

$$PV(K) = \sum_{k=0}^{\infty} \frac{R_k}{(1+r)^k} \qquad (3.59)$$

where r is the annual discount rate and R_k is operating revenue in year k. Thus

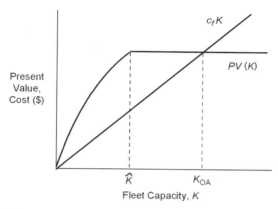

Figure 3.9 Present value of net fleet operating revenue, and total cost of the fleet, for the model of regulated open access.

$$PV(K) = \bar{R} \cdot \frac{1+r}{r} \quad \text{if } K \geq \hat{K} \tag{3.60}$$

Again this is a constant, independent of fleet size K. The graph of $PV(K)$ is shown in Fig. 3.9. (The nonlinearity of $PV(K)$ for $K < \hat{K}$ follows from the fact that if $K < \hat{K}$ then the annual quota Q is not caught, so that the biomass x gradually increases over time. The factor $(pqx - c)$ in Eq. (3.58) varies from year to year, and is a decreasing function of K. Thus the curve $PV(K)$ in Eq. (3.59) has decreasing slope as K increases, for $K < \hat{K}$.)

Assuming instant (pulse) investment at time $t = 0$, we see that in the absence of any control over fleet capacity, we have

$$PV(K_{OA}) = c_f K_{OA} \tag{3.61}$$

where K_{OA} denotes the open-access fleet size. (If some vessels are already active in the fishery at time $t = 0$, Eq. (3.61) predicts how far expansion will proceed. If current capacity exceeds K_{OA}, there will be no additional expansion.) From Eq. (3.60) this implies that

$$K_{OA} = B/c_f \quad \text{if } c_f \leq B/\hat{K} \tag{3.62}$$

where

$$B = \bar{R}\frac{1+r}{r} \tag{3.63}$$

(see Fig. 3.9), i.e., B equals the present value of fleet operating revenues.

If $c_f > B/\hat{K}$ then $K_{OA} < \hat{K}$, and possibly $K_{OA} = 0$—sufficiently high capital costs may cause the fishery to be non-viable, as in all our models. This is not the case of interest here, however.

Let us continue with the numerical example, with

$$p = \$1,000/\text{tonne}$$
$$c = \$150,000/\text{SVU year}$$
$$c_f = \$500,000/\text{SVU}$$
$$r = 0.1$$

Thus $R = \$7$ million and $B = \$77$ million (of course the present value B in Eq. (3.63) depends strongly on the discount rate r). Thus $K_{OA} = 154$ vessels. Excess capacity under regulated open access is 114 vessels, and K_{OA} is nearly four times the optimal capacity $\hat{K}$.

The amount of excess capacity in the example depends on parameter values. For example, by Eq. (3.62), total fleet size K_{OA} is inversely proportional to fixed vessel cost c_f. If $c_f = 1$ million we get $K_{OA} = 77$, so excess capacity is 37 vessels. Or if $c_f = \$1.9$ million then $K_{OA} = 40$, so there is apparently no excess capacity. (But consider Fig. 3.9 for this case. Now the cost line intersects the PV curve at $K = \hat{K}$. Because of the curvature to the left of $\hat{K}$, the optimal fleet capacity is $K^* < \hat{K}$. hence $K_{OA} = 40$ is again larger than the optimum.)

Excess capacity in the present equilibrium model of a regulated open-access fishery results entirely from "divvying up" the fixed reward B. It costs c_f per vessel to enter the game, so the predicted number of entrants is B/c_f. In reality, entry to a restricted open-access fishery may be larger than predicted by the present model. For example, if the initial stock x_0 is larger than x (the managed equilibrium), then larger initial revenues will attract additional capacity (as shown in Sec. 3.2). Also, if vessels in the given fishery transfer from other fisheries, either on a seasonal or a permanent basis, the fixed cost c_f will be smaller than the original purchase costs of a vessel, and entry will be larger than otherwise predicted. Finally, the anticipation of increased future revenues may also trigger further expansion of fleet capacity. The prospect that participation will be limited at some future time, for example, may encourage an expansion of current capacity. This will be an important consideration in our analysis of buy-backs.

Note also in passing that the level of excess capacity in a regulated fishery could in fact be greater than the level of excess capacity that would occur in the absence of catch regulation. Such regulation maintains profitability of the fishery in terms of operating revenues, which would otherwise tend to be dissipated because of overfishing. In Canada's marine fisheries, for example, it seems certain that government regulation of annual catches, combined with generous subsidization of the fishing industry, has contributed substantially to the current crisis of overcapacity. Repeated attempts to reduce the size of salmon and other fleets through buy-back programs appear to have had little success (Holland et al. 1999). We next turn to the economic theory of buy-backs.

Buy-Back Programs

Consider a buy-back program designed to reduce fleet capacity to the optimal level (e.g., 40 vessels) by removing excess capacity (114 vessels) permanently. The managers announce a buy-back price of p_b per standard vessel (scaled according to fishing capacity for actual vessels). Acceptance of the buy-back payment is voluntary, but imposes an obligation to remove the vessel irrevocably from the fishery. (What happens to the removed vessels does not concern us here.)

It is reasonable to assume that a vessel owner will withdraw from the fishery only if the buy-back payment p_b is at least as large as his expected present value from remaining in the fishery. Assuming that K' vessels remain in the fishery, each remaining owner will have $PV' = PV(K')/K' = B/K'$. If $p_b > B/K'$ additional vessels will accept the payments and withdraw, until

$$K' = B/p_b \tag{3.64}$$

Thus, in order to achieve the optimal fleet size $K' = \hat{K}$ we require that

$$p_b = B/\hat{K} \tag{3.65}$$

In the example, this gives $p_b = \$77$ million $\div$ 40 = \$1.925 million per vessel.

Can this be correct? Does it really take a payment of almost \$2 million to induce sufficiently many owners to retire their vessels? This is almost four times the original cost of a vessel.

First, why is $p_b > c_f$? Why does the fishery suddenly become so valuable that vessel owners expect future revenues to be worth \$1.9 million per vessel, in terms of present value? The answer is that the proposed limited entry program will result in annual operating revenues per vessel equal to $B/\hat{K}$, which equals p_b. By assumption $\hat{K} < K_{OA}$ (Fig. 3.9), and $K_{OA} = B/c_f$ by Eq. (3.62). It follows that $p_b > c_f$.

In words, at open-access equilibrium (K_{OA}), the present value of future per vessel operating revenues just match vessel cost (c_f), whereas under optimal limited entry, per-vessel revenues will be much larger. Owners will not voluntarily withdraw unless they receive comparable compensation. It is very important to understand this argument, which has seldom been mentioned in the literature on buy-back programs. To be successful, a voluntary buy-back scheme must offer large buy-back payments, possibly much larger than the actual value of a vessel. This argument indicates why most actual buy-back programs have failed to remove substantial amounts of fishing capacity. Presumably governments have been reluctant to offer buy-back payments that greatly exceed the original cost of vessels. In our example, the total cost of the buy-back program would be $114p_b = \$219$ million, compared to the present value of \$77 million of the fishery.

The actual effectiveness of real-world buy-back programs may be lower than suggested by our model. A buy-back program may only entice the withdrawal of the least efficient vessels. For example, in Canada's Pacific salmon fisheries, three successive buy-back programs appear to have had little effect on overall fleet capacity (James 2004).

Next we argue that a buy-back program may actually induce an *increase* in fishing capacity if fishermen anticipate that such a program will be instituted. Such anticipatory increases in capacity have in fact frequently occurred. The reason is straightforward: in order to qualify for a buy-back payment (or for a license in the limited-entry fishery), one must be a participant in the fishery when the program is initiated. According to our analysis, the rewards will exceed the cost of a vessel (otherwise the program will not be successful). If vessel owners anticipate a future buy-back, they will be more strongly motivated to enter the fishery than they would be in the absence of buy-backs (Clark et al. 2005).

The conclusion to be drawn from this very simple analysis is that *voluntary buy-back programs are exceedingly dangerous*. They have little prospect of success unless government expenditure on buy-back payments greatly exceeds the economic value of the fishery. In addition, they can actually induce further expansion of overcapacity—quite the opposite of the intended objective. We next discuss an alternative approach.

Mandatory, Fishery-Financed Buy-Backs

The prediction that an effective buy-back program may be unacceptably costly depends in part on the assumption that the acceptance of buy-back payments is voluntary. Vessel owners will not accept a buy-back payment that is lower than the benefits they can expect to earn by remaining in the fishery. An alternative to voluntary buy-backs would be mandatory buy-backs, financed by taxes, or royalties, paid by the remaining licensed vessel owners.

To implement such a buy-back program, each existing vessel owner would be given the option of accepting a buy-back payment of some specified amount x, or receiving a license to continue fishing and paying a license fee y. (The license fee could theoretically be replaced by a levy on catches over future years.) In order for this system to be self-financing, we require that

$$(K_{\text{OA}} - \hat{K})x = \hat{K}y$$

Also, for equity between those who are bought out and those who remain, we need

$$x = \frac{B}{\hat{K}} - y$$

Solving for x and y gives

$$x = \frac{B}{K_{\text{OA}}} = c_f \qquad (3.66)$$

by Eq. (3.62), and

$$y = \frac{B}{\hat{K}} - \frac{B}{K_{\text{OA}}} \qquad (3.67)$$

For the numerical example, this gives a buy-back price of $500,000 per vessel (i.e. the original cost of the vessel), and a one-time license fee of $1.425 million per vessel. The net present value of one license is then also $500,000.

In principle, given these values of x and y, vessel owners should be indifferent between leaving and staying in the fishery. Exactly how it would be decided who stays and who leaves is not our concern; possibly some kind of lottery would be used.

But why does it turn out that the buy-back price x is exactly equal to the vessel cost c_f? This is a consequence of our perfect foresight assumption that under open access PV per vessel exactly equals the cost per vessel, i.e. $B/K_{\text{OA}} = c_f$. Under the limited entry program, total PV is again divided evenly among all original vessel owners, so that the buy-back price equals B/K_{OA} also. Hence buy-back price equals vessel cost.

But what if the perfect foresight assumption is not correct? For example, the actual number of vessel units K_{actual} might exceed B/c_f for various reasons: $K_{actual} > B/c_f$. If we still assume that the buy-back program divides the resource value B equally among all existing vessel units, then the buy-back price will equal $B/K_{actual} < c_f$. In this case the fishery is not sufficiently profitable to completely recompense all original vessel costs.

Implications and Limitations of the Analysis

Two aspects of the foregoing analysis seem robust and significant in practical terms:

1. To be successful in terms of reducing fishing capacity, a voluntary buy-back program is likely to be extremely expensive. A program that merely offers to buy up vessels at their original cost may have little actual effect on fleet capacity.
2. A voluntary buy-back program, if anticipated by potential vessel owners, may precipitate a substantial increase in fleet capacity.

A mandatory, self-financing buy-back program might seem to avoid these problems. However, it is not clear whether such a program is feasible in practice. Those vessel owners who obtain licenses would be required to pay large fees or royalties to compensate owners who give up their vessels. Such fees would doubtlessly be unpopular. In fact, receiving a buy-back payment might be considered a better option than continuing to fish under such a fee system. If so, few owners would be willing to participate, and the scheme would collapse.

We seem to have reached an impasse. Both voluntary and mandatory buy-back programs have sufficiently serious drawbacks to imply that they cannot succeed. Indeed, as explained shortly, it is not even clear from our models that a "successful" buy-back program would actually achieve anything worthwhile. Before addressing this question, let me point out here that what is missing from the analysis so far is the possibility of somehow altering the economic incentives of fishermen. It will be argued in Chapter 4 that the use of IFQs (individual fishing quotas) can alter fishermen's incentives in a useful way. Indeed, an IFQ program, if properly designed, monitored, and enforced, may of itself remove the incentive for overcapacity, and also eliminate the need for a buy-back program. However, there may also be severe anticipatory problems associated with IFQs. More on this in Chapter 4.

Returning to buy-back programs without IFQs, we now ask what benefits would be realized from a program that actually succeeded in reducing the original capacity to the desired level $\hat{K}$? According to our model, nothing! The annual catch would remain the same, total annual revenues would remain the same, and fishing costs would remain the same. Fewer vessels would be operating over a longer fishing season, but this would not alter either benefits or costs.

Surely something must be missing from the model, but what? What do its proponents claim will be the benefits of fleet reduction? Here is a partial list:

(a) prevent economic waste
(b) prevent overfishing
(c) improve manageability of the fishery
(d) reduce political pressure for large or increased catch quotas
(e) improve the quality of the catch
(f) reduce the degree of danger associated with fishing activities

Although most of these benefits might in fact occur, most would be doubtful unless economic incentives are also altered. We will argue in Chapter 4 that all these benefits can occur under an IFQ system.

Here let us consider the question of economic waste. According to our model, economic waste would not be reduced at all by a buy-back program. The dictum "excess capacity makes economic waste" may seem to be a truism (which it is), but it does not follow that removing excess capacity will eliminate the waste. The waste could have been prevented if the excess vessels had never been built, but buy-back programs are by definition concerned with removing, not preventing excess capacity.

Are there other sources of economic waste that would in fact be reduced as a result of vessel buy-backs? Yes: mobilization costs. Such costs were not considered in the above models, but let us suppose now that each vessel participating in the fishery incurs an annual mobilization cost c_m. This cost might be associated with traveling to and from the fishing ground, or with preparing the vessel for sea. The total fleet annual mobilization cost $c_m K$ would be reduced if the number of active vessels K were to be reduced. Fleet reduction can reduce waste by reducing mobilization costs.

Another possible example of economic waste, or inefficiency, resulting from excess capacity is interference caused by undue crowding of vessels.

If crowding occurs, the catch rate would be some nonlinear function of fleet size K, rather than the linear function $qxDK$ assumed in our model. If so, reducing fleet capacity would improve efficiency.

Thus in general it is possible that fleet reduction could reduce economic waste, not by saving on the capital costs of vessels, but by improving efficiency of fishing activities. However, since the conditions conducive to a derby fishery still prevail, efficiency savings might be quite minimal. Regarding conservation objectives, it is possible that significant fleet reduction would reduce the pressure on fish stocks, and allow managers to set conservative, precautionary catch quotas. However, as we have seen, both theory and experience suggest that the quest for improved conservation may be thwarted because of basic economic incentives—unless these incentives are changed. Experience (and theory—see Ch. 4) also suggests that incentive alteration can be achieved through the use of IFQs (individual fishing quotas). IFQ systems have their own problems, of course, but it is likely that these can be overcome with sufficient care in the design and implementation of the system. An IFQ system may automatically result in fleet reduction, without any need for government financial assistance. For these reasons, IFQs will probably become the core of future fisheries management programs. The importance of substantial, rent-capturing fees for individual quotas also needs to be emphasized. Also, effort quotas may be preferable to catch quotas in some circumstances, including multispecies fisheries.

Multipurpose Fleets and Spillover Effects

Our models of optimal and open-access fishing capacity pertain to the situation where a fleet of identical, or at least similar, specialized vessels exploits a single stock of fish. While vessel specialization is not uncommon, many actual fishing vessels do participate in more than one fishery. Two contrasting cases can be identified; multipurpose vessels that switch target species on a seasonal basis, on the one hand, and vessels that switch permanently to new stocks when existing stocks are depleted, on the other hand.

The theory developed in this chapter can be extended to deal with such cases, but we will not pause to do so here. One thing is fairly clear, however: the capital cost c_f for a vessel that enters a fishery after having fished elsewhere will normally be much smaller than the original cost of that vessel. In fact c_f may be close to zero unless there are substantial costs of refitting. In this case extreme overcapacity may result from such "spillovers" of previously used fishing vessels. For example, consider the case in which the spillover consists of vessels that have become available as the result of a buy-back program in some other fishery. In this event, the buy-back program will merely have the effect of transferring overcapacity from one fishery to another. For this reason many fleet reduction programs have required physical destruction of vessels that receive buy-back payments.

Capital Stuffing

Finally we briefly consider the problem of capital stuffing. (See Sec. 4.2 for a detailed model.) This term refers to the upgrading of licensed vessels in a limited-entry fishery (Weninger and McConnell 2000). The possibilities are numerous: replacement of an existing vessel by a larger, more powerful vessel (unless this is prohibited); addition or expansion of freezer capacity; addition of electronic gear for navigation or for finding fish; and so on. All such modifications are intended to increase the vessel's fishing power, thereby increasing its share of the total catch.

Capital stuffing tends to defeat the aims of the limited-entry program, by increasing the effective capacity of the fleet. Also, unless the measure of fishing effort is recalibrated to account for such capital stuffing, the CPUE index of stock abundance will become progressively more inaccurate, and will increasingly overestimate stock abundance.

But can't capital stuffing be controlled through rigorous management? Perhaps, but doing so has proven to be difficult in practice. Regulations are required that cover virtually every aspect of the vessel, such as length, width, tonnage, engine HP, cargo capacity, ancillary equipment, and so on. Though common nowadays in limited-entry fisheries, such regulations are not always sufficient to thwart capital stuffing, which emanates from the economic incentives of the fishermen.

What are these economic incentives? After all, if each vessel owner carries out capital stuffing to the same extent, no one will increase their catches—assuming that managers are able to maintain the TAC. Indeed, net revenues will actually decline, as a result of the money spent on capital stuffing. Surely this folly will be recognized and avoided.

Not so. Box 3.3 uses a game-theoretic model to explain why the apparently irrational phenomenon of capital stuffing occurs. In a nutshell, any vessel owner who singlehandedly stuffs his vessel can increase his share of the TAC provided other owners don't do the same thing. Likewise, any owner who fails to stuff will lose out if other owners do it. An equilibrium is reached only when

Box 3.3. A game-theoretic model of capital stuffing.

The theory of games was developed in the 1950s as a method of understanding economic behavior in conflict situations (von Neumann and Morgenstern 1947; Nash 1951, 1953). Game theory has since become a central paradigm in economics.

Here we consider a simple 2×2-matrix model for capital stuffing. First, imagine that there are two vessel owners, A and B, the only participants in a TAC-regulated fishery. Each has the option to stuff or not to stuff. The payoff matrix is

		B	
		stuff	not stuff
	stuff	(1,1)	(3,0)
A			
	not stuff	(0,3)	(2,2)

The entries (x, y) are the net revenues to A and B respectively. If both owners refrain from stuffing, they share a total net revenue of 4 units (say, \$4 million). If A stuffs and B doesn't, then A takes the entire TAC, but pays one unit stuffing cost, so the payoffs are (3,0). Finally, if both owners stuff, they share the TAC equally, but each pays the stuffing cost. This simplistic model qualitatively captures the essence of the capital stuffing problem.

What is the "solution" to this game? Consider A's decision. If B chooses to stuff, then A should stuff, too (payoff 1 versus 0). And if B does not stuff, A still does better by stuffing (3 versus 2). By symmetry, both owners do best by stuffing, and this is the so-called competitive solution to this capital stuffing game (Nash 1951). The competitive solution contrasts to the cooperative solution in which both owners refrain from stuffing (Nash 1953). (The above game structure is known in the literature as the prisoners' dilemma.)

Clearly A and B would benefit by cooperation—provided they trust each other. But there is always a temptation to cheat, even though cheating invites retaliation. A large literature exists on the topic of repeated two-person prisoners' dilemma games (see

---

Box 3.3 *continued*

(Dugatkin 1997). The general consensus seems to be that
cooperation is the most likely result, but experiments with animal
subjects have not always confirmed this prediction (Stephens et al.
2002; Mesterton-Gibbons and Adams 2002).

Limited-entry fisheries usually involve N owners, $N > 2$. As a
start, let us imagine that the $N = 6$ owners agree not to use capital
stuffing, but that a certain owner (A) contemplates breaking his
promise. The payoff matrix is

		Coalition of 5 owners	
		stuff	not stuff
Defector A	stuff	(3,15)	(5,18)
	not stuff	(0,19)	(4,20)

where the payoffs (x, y) are received by A and the remaining
coalition of 5 owners respectively. Here, the total payoff with no
stuffing is 24 units, i.e. 4 units per owner. If A stuffs and the other
owners do not, then A gets 6 units of catch but pays a one unit
stuffing cost. The remaining coalition gets $24 - 6 = 18$ units. The
other entries are calculated in the same way. The competitive
solution here is a bit different from the two-owner game. Here the
defector A improves his payoff by stuffing, regardless of what the
coalition does, but the coalition does best by not stuffing,
regardless of what A does. However, the coalition does lose 2 units
to the defector. Both the defector and the coalition would do best if
everyone adhered to the non-stuffing agreement.

(What about the idea of perfect foresight in this setting? The
defector predicts that the coalition will not stuff, and he stuffs. The
coalition doesn't stuff, even though it predicts that the defector
will. Thus the competitive solution satisfies the definition of perfect
foresight—assuming that the residual coalition holds to the
agreement.)

---

further stuffing is more costly than the perceived benefits. Thus profit
(or rent) dissipation, which our model assumes will be eliminated by

a limited-entry program, may in fact re-emerge as a result of capital stuffing. Holland et al. (1999) attribute the failure of many buy-back programs to this phenomenon.

How then can capital stuffing be prevented? If each vessel owner has a fixed catch quota (an IFQ), then the economic incentive for capital stuffing is eliminated, since nothing can be gained by stuffing. There may be an incentive to adopt improvements that reduce costs, but this does not constitute capital stuffing. But to be effective, the individual quotas must be rigorously enforced. We discuss IFQs in more detail in Chapter 4.

3.5 Summary of Chapter 3

This chapter is concerned with fishermen's investment decisions (i.e., purchasing a vessel) and their consequences. As in Chapter 2 we consider both open-access and optimally controlled fishery models. Technically speaking, the models are state-variable models with two state variables; the population biomass $x(t)$ and fleet effort capacity $K(t)$. Model equations are:

$$\frac{dx}{dt} = G(x) - qEx \qquad (3S-1)$$

$$\frac{dK}{dt} = I - \gamma K \qquad (3S-2)$$

$$0 \leq E(t) \leq K(t) \qquad (3S-3)$$

$$0 \leq I(t) \qquad (3S-4)$$

Here $K(t)$ denotes fleet capacity, i.e. maximum effort capacity. Operating revenue $R(t)$ is

$$R(t) = (pqx(t) - c)E(t) \qquad (3S-5)$$

The main feature of this model is the assumption of irreversibility in fleet capacity, Eq. (3S − 4).(We also briefly considered the possibility of quasi-reversibility, allowing for the sale of excess vessels at a reduced price.)

Irreversibility (or non-malleability) of vessel capital is what underlies the problem of overcapacity. Fishing activity that may have been highly profitable during the initial stage of the fishery is no longer so profitable after the stock has been fished down. For example, at open-access bionomic equilibrium $x = x_{OA}$, operating revenues $R(t)$ are zero, and vessel owners thereafter obtain zero return on their original investment. Positive returns are made during the initial phase of the fishery, however. Indeed, it is the prospect of large initial profits that attracts vessels to the fishery in the first place. In our analysis we assumed that potential vessel owners had perfect foresight (sometimes called "rational expectations"), in the sense that they could correctly forecast the future of the fishery, and make their investment decisions accordingly. While this assumption may be extreme, the alternative assumption that vessel purchasers ignore future developments is also unsatisfactory. Any experienced fisherman knows that the initial profitability from a new fishery is unlikely to last indefinitely. (If one insists on assuming myopic decision making, so that fishermen assume that current catch rates will continue unchanged, the predicted level of overcapacity will be greater than that deduced from our model.)

Our model thus predicts that in the absence of any regulation, a newly developing fishery will experience both bioeconomic overfishing and overcapacity.

Next, we consider the case of regulated open access, with TAC-based regulation aimed at maintaining some specified stock level x_1. How x_1 is determined is not our concern here, but we do assume that $x_1 > x_{OA}$. The effect of this form of regulation depends on the state (x, K) of the fishery at the time that regulation is introduced, and also on whether the regulation is anticipated in advance. If the TAC-based regulation is anticipated before the stock has been severely reduced, the result may be additional overcapacity, motivated by the realization that future operating revenues $R = (pqx_1 - c)E$ will never decrease to zero. If the TAC regulation is not anticipated, fishing capacity may either increase or remain unchanged after the regulation is introduced. In most cases the TAC-regulated fishery will experience excess fishing capacity.

It therefore seems desirable that fleet capacity should be controlled to prevent or eliminate overcapacity. How can this be accomplished? Limiting capacity at the outset in a new fishery is seldom attempted, and would be difficult because of lack of information. The usual situation is a fishery that already involves excess capacity, which could then in principle be reduced by means of a vessel buy-back program.

Unfortunately, buy-back programs have often failed to achieve their objectives, for two reasons that are both related to anticipation. First, if vessel withdrawal is voluntary, fishermen who opt to sell their vessels and retire will expect to receive compensation equivalent to the average future profits of those who remain in the fishery. An effective buy-back program may therefore be extremely costly. In many cases only the least efficient vessels and operators may withdraw from the fishery.

Second, if a future buy-back program is anticipated in advance, the result will be a greater level of capacity buildup than would otherwise occur. Buy-back programs may therefore result in greater economic losses than would occur without them.

A further difficulty with buy-back programs is that, like other limited-entry programs, they often induce capital stuffing. Given that the limited fishing fleet competes for the TAC, vessel owners are motivated to try to increase the fishing power of their vessels. Thus capital tends to seep back into the fishery; controlling this process has often proved to be difficult.

To summarize, management strategies based on such methods as gear restrictions, TACs, limited entry and vessel buy-backs do not directly change the economic incentives of fishermen. Under such programs individual fishermen continue to participate in a non-cooperative, "prisoners' dilemma" type of game, trying to maximize their share of a finite total catch quota. What is needed instead is a management approach that does alter economic incentives in a way that fosters cooperative behavior among the fishermen. One such method, based on individually allocated quotas, will be studied in detail in Chapter 4.

4

Fisheries Management

Fisheries management techniques can be divided into two classes, those that do not specifically address the open-access aspect of the fishery, and those that do. The first class includes the traditional methods of total catch quotas (TACs), area and seasonal closures, gear regulations, and the like. These methods can succeed in preventing overfishing and resource depletion, but they usually also lead to excess fishing capacity, for reasons set out in Chapter 3. Attempts to reduce overcapacity through license limitation and buy-back programs often encounter difficulties as discussed also in Chapter 3.

In this chapter we discuss management methods that directly counter the open-access problem. Two approaches that will be analyzed are first, taxes or royalties on the catch (or on effort), and second, individual catch (or effort) quotas. These can also be used in combination. We will use the term "royalties," rather than "taxes," to stress the point that these charges pertain to the right to exploit a publicly owned resource. (Besides, everyone abhors taxes, but charging royalties, or user fees, for the right to harvest a publicly owned resource may seem fair to most people.) Both methods—royalties and individual quotas—alter the economic incentives of fishermen in a way that promotes both economic efficiency and resource conservation. Clearly, however, royalties and individual quotas have opposite distributional implications. Royalties capture resource profits for the government, whereas individual quotas (if freely awarded) grant profits to the quota recipients. Not surprisingly, the fishing industry has exhibited a preference for individual quotas, often arguing that royalties would be unfair.

Many analysts have argued that the choice of royalties versus individual quotas is purely a question of equity, with little relevance to the basic problem of achieving economic (and biological) rationalization of commercial fisheries. We will challenge this claim, and show that because of the possibility of anticipation (rational expectations), rent-garnering royalties are essential for achieving positive economic benefits from marine resources. The ideal management system employs individual quotas and substantial royalties in combination—precisely the same approach that has been widely used in other resource industries.

4.1 Royalties

Resource royalties garner some or all of the resource rent for the public purse. For countries with resource-based economies, resource royalties are often a major source of government revenue. This has seldom been the case, however, for marine fishery resources. Historically most of these resources were unregulated, so that sustained profits were nil, although substantial temporary profits may have been obtained during the initial development of the fishery. The possibility of preserving and capturing sustained profits from fisheries has existed since the establishment of EEZs (Extended Economic Zones) in the 1970s, but with some exceptions this opportunity has not generally been taken advantage of.

The Basic Model

We first consider the basic dynamic model of Chapter 2:

$$\frac{dx}{dt} = G(x) - h, \quad x(0) = x_0 \tag{4.1}$$

$$h = qEx \tag{4.2}$$

$$R = ph - cE = (pqx - c)E \tag{4.3}$$

Our model of the unregulated open-access fishery is characterized by

$$E(t) = \begin{cases} E_{\max} & \text{if } x(t) > x_{\mathrm{OA}} \\ G(x_{\mathrm{OA}})/qx_{\mathrm{OA}} & \text{if } x(t) = x_{\mathrm{OA}} \\ 0 & \text{if } x(t) < x_{\mathrm{OA}} \end{cases} \tag{4.4}$$

where $x_{\mathrm{OA}} = c/pq$ is the bionomic equilibrium biomass.

Suppose now that a royalty τ is imposed on all harvested fish. Assume first that $\tau = $ constant. Then the net after-royalty revenue flow is

$$R_\tau = (p - \tau)h - cE = ((p - \tau)qx - c)E \tag{4.5}$$

Thus effort is again given by Eq. (4.4), but with x_{OA} replaced by x_τ where

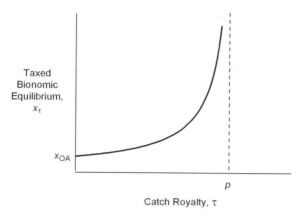

Figure 4.1 Bionomic equilibrium x_τ as a function of the catch royalty τ.

$$x_\tau = \frac{c}{(p - \tau)q} \qquad (4.6)$$

Note that any desired equilibrium biomass level $x_\tau > x_{OA}$ can be achieved by using an appropriate catch royalty τ, with $0 < \tau < p$ (see Fig. 4.1). (Indeed, it would even be possible to shut down the fishery entirely, with a sufficiently high royalty τ, such that $x_\tau > K$, the carrying capacity of the population.)

In particular, by determining τ so that $x_\tau = x_\delta$ (the optimal equilibrium biomass—see Eq. 2.63), the managers would theoretically be able to achieve this optimal equilibrium without any further management actions. Moreover, imposing the constant royalty τ would result in a dynamically optimal fishery—at least according to our present model. This follows from noting that Eqs. (4.4) still hold under the royalty, but with x_{OA} replaced by $x_\tau = x_\delta$. As shown in Chapter 2, Eq. (2.62), this is the dynamically optimal fishing strategy for the basic model.

This is an appealing result—a single, constant landings royalty is sufficient to force the otherwise open-access fishery into the dynamically optimal exploitation mode. Unfortunately, this result is strongly model-dependent, as we shall see.

Note that if $x(0) > x_\delta$ the constant-royalty policy garners some but not all of the profits from the initial stock-reduction phase. Here we have $R_\tau > 0$, so some of the profits are captured by the fishermen. However, if we also consider fleet capacity, these transitory residual profits would tend to be dissipated through overcapacity, although this would be reduced because of the royalties.

Information and Controllability

Let x now denote the target biomass level chosen by fishery managers; for example we might have $x = x_{\mathrm{MSY}}$ or $x = x_\delta$, or some other specification. Solving Eq. (4.6) for the royalty τ, we have

$$\tau = p - \frac{c}{qx} \qquad (4.7)$$

The royalty rate that (with no other type of control) would be required to achieve the target x thus depends on three parameters p, c and q. Any error in measuring these parameters would be reflected in an error in τ and hence also in the bionomic equilibrium x_τ, so that $x_\tau \neq x$. Suppose, for example, that price p, though initially assessed correctly, subsequently increases over time, but that the catch royalty τ remains at its initial level. The result would be progressive overfishing as x_τ steadily decreases. The same would occur if technological changes resulted in a decrease in effort cost c.

A second difficulty arises from the possibility of model error. For example, the above model assumes a Type II catch-effort relationship $h = qEx$. If the correct model is Type I, with $h = qEx^\beta$ ($\beta < 1$), the effect of catch royalties on the stock level x_τ will be weaker than predicted. In the extreme case with $\beta = 0$, a royalty on catch would either have no effect on fishing (if $\tau < p - c/q$) or else would shut down the fishery entirely.

In other words, the extent to which a royalty on catches is able to control an otherwise unregulated fishery depends on the form of the catch–effort relation. Calculating the royalty rate that would achieve a particular management objective, and adjusting this rate over time to follow changing circumstances, could be difficult indeed.

The conclusion to be drawn from this discussion is that catch royalties are not a reliable instrument, by themselves, for controlling the exploitation of a fishery. Royalties should therefore be used in conjunction with direct harvest controls such as TACs.

Effort Royalties

Royalties on fishing effort are an alternative to catch royalties. Effort royalties, or fees, may be simpler to administer than catch royalties, but this simplicity may be offset if fishermen can more easily "cheat" on their effort level than on their catches. Here we will ignore this important possibility, however.

If τ_E denotes the effort royalty (\$/SVU yr, for example), we obtain

$$R = ph - (c + \tau_E)E = (pqx - (c + \tau_E))E$$

The bionomic equilibrium is now given by

$$x_{\tau_E} = \frac{c + \tau_E}{pq} \tag{4.8}$$

Thus the effort royalty τ_E required to achieve bionomic equilibrium at a specified stock level x is

$$\tau_E = pqx - c \tag{4.9}$$

The previous discussion of controllability for catch royalties also applies to effort royalties. In addition, annual effort may be more difficult to assess accurately than annual catches. Nevertheless effort royalties can extract a large portion of resource profits, to the advantage of the public purse.

Royalties Plus TACs—The Taxed Derby Fishery

A fishery managed solely on the basis of annual total catch quotas (TACs) typically develops into a derby fishery, in which each participant exerts an unduly high daily level of effort during the brief seasonal opening (see Sec. 2.6) occasioned by the existence of excess fishing capacity. How would a catch royalty affect this outcome?

Recall that under derby conditions each fisherman attempts to maximize his daily net revenue:

$$\text{maximize } (pqxE - c(E)) \tag{4.10}$$

where E denotes daily effort. Except for corner solutions (see Eq. 2.82) this implies that $E = E_0$ where

$$c'(E_0) = pqx \tag{4.11}$$

Though individually optimal, the effort level E_0 is greater than the fishery-rent-maximizing level E^*. This is the feature that, together with excess capacity, characterizes the derby fishery.

Now suppose that fishermen are required to pay a catch royalty τ. Then Eq. (4.10) becomes

$$\text{maximize } ((p - \tau)qxE - c(E)) \tag{4.12}$$

and this implies that $E = E_\tau$ where

$$c'(E_\tau) = (p - \tau)qx \tag{4.13}$$

Thus by setting an appropriate royalty level τ^*, given by

$$\tau^* = p - c'(E^*)/qx \tag{4.14}$$

the managers can in principle induce fishermen to employ the optimal effort level $E_\tau = E^*$. Also, by reducing profitability in the fishery, the catch royalty will reduce the incentive for overcapacity.

As before, accurate specification of the royalty τ^* may not be possible, but any royalty that captures a substantial portion of the profits will have an economically beneficial effect on the fishery. Since catch rates are assumed to be controlled by means of TACs, mis-specification of the royalty will have no effect on resource conservation, unless the royalty is so large as to discourage fishing altogether.

How would a fee on effort be represented in this model? The managers can probably not assess each vessel's daily effort E, but perhaps they could impose a daily fee ϕ for each day spent fishing. This would add the constant ϕ to the daily cost function:

$$c_\phi(E) = c(E) + \phi \tag{4.15}$$

This in turn increases the shut-down effort level $E_{1,\phi}$ defined by

$$c'_\phi(E_{1,\phi}) = c_\phi(E_{1,\phi})/E_{1,\phi} \tag{4.16}$$

—see Fig. 4.2. Thus the daily fee will either have no effect on an individual's daily effort E_0, or it will cause the individual to cease fishing entirely (if $E_{1,\phi} > E_0$). (If the fleet contains vessels with different cost functions, the high-cost vessels would be eliminated.)

Fleet Capacity

A royalty on catch (or effort), by reducing the profitability of fishing, will also reduce the incentive for vessels to enter the fishery. For example, consider the case of an otherwise unregulated open-access fishery, subject to a catch royalty τ. Using the notation of Chapter 3, let fleet operating revenue, under royalties, be $R_\tau(t)$:

$$R_\tau(t) = ((p - \tau)qx(t) - c)E(t) \tag{4.17}$$

where

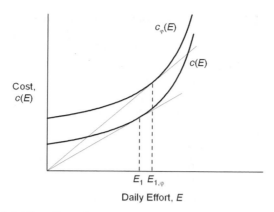

Figure 4.2 The effect of a daily fishing fee ϕ is to raise the shut-down effort level $E_{1,\phi}$.

$$E(t) = \begin{cases} E_{\max} & \text{if } x(t) > x_\tau \\ G(x_\tau)/qx_\tau & \text{if } x(t) = x_\tau \\ 0 & \text{if } x(t) < x_\tau \end{cases} \qquad (4.18)$$

where the bionomic equilibrium x_τ is given by Eq. (4.6). We also have

$$PV_\tau(E_{\max}) = \int_0^\infty e^{-\delta t} R_\tau(t)\, dt \qquad (4.19)$$

Clearly $PV_\tau(E_{\max})$ decreases as the royalty level τ is increased, and this implies that the initial fleet capacity $E_{\max,\tau}(0)$ also decreases as τ is increased.

More generally, the state-space diagram of Fig. 3.4 is altered in the obvious way, when catch royalties are introduced. Namely, both stock thresholds x_{OA} and x_{I} shift to the right, and the investment curve S shifts right and down, as τ is increased. Thus catch royalties shift the otherwise unregulated fishery in the direction of the optimum, Fig. 3.7. The question then arises whether there is an optimal constant catch royalty τ^* that actually achieves the full dynamic, harvest-plus-investment, optimum. The answer is negative—a second royalty (or possible a subsidy) on investment is required. But this is probably far too sophisticated for practical application. In a fishery managed with TACs, a single catch royalty should capture a large portion of the profits, and control over-capacity.

4.2 The Limited-Entry Fishery

We next use our model of a TAC-regulated fishery (Sec. 3.4), making the additional assumption that participation is limited to a given set of N firms. Each firm can employ any effort level, up to its capacity $E_{\max}$. To begin with, firm capacity $E_{\max}$ is treated as an unrestricted decision variable. We first assume, for simplicity, that the N firms are identical in terms of economic parameters.

Total annual fleet harvest H is given by

$$H = \min(qxNE_{\max}D_1, Q) \tag{4.20}$$

where the symbols have similar interpretations to those in Eqs. (3.49)–(3.55). Specifically

$$H = \text{annual harvest}$$
$$D_1 = \text{maximum season length } (0 < D_1 \le 1)$$
$$Q = \text{annual quota or TAC}$$

The optimal total fleet capacity is $\hat{K} = Q/qxD_1$, so the optimal capacity per firm $\hat{E}_{\max}$ equals $\hat{K}/N$, or

$$\hat{E}_{\max} = \frac{Q}{NqxD_1} \tag{4.21}$$

For example, suppose $N = 5$ firms, and adopt the numerical example of Sec. 3.4. Then $\hat{K} = 40$ SVU, so that

$$\hat{E}_{\max} = \frac{40}{5} = 8 \text{ SVU}$$

This is the optimum effort capacity per firm.

Net fleet annual operating revenue is (if $E_{\max} \ge \hat{E}_{\max}$)

$$R = pQ - cNE_{\max}D$$
$$= \left(p - \frac{c}{qx}\right)Q \tag{4.22}$$

Also, the present value of fleet operating revenues is

$$PV = R\left(\frac{1+r}{r}\right) \tag{4.23}$$

Present value per firm is PV/N.

Under the regulated open-access model, firms will expand their capacity until $PV = c_f NE_{\max}$. Thus the open-access capacity per firm equals

$$E_{\text{max,OA}} = \frac{PV}{Nc_f} = \frac{1}{Nc_f}\left(p - \frac{c}{qx}\right)Q\left(\frac{1+r}{r}\right) \qquad (4.24)$$

where PV is total fleet present value of operating revenues, as in Eq. (4.23). For the numerical example this gives

$$E_{\text{max,OA}} = \frac{154}{5} \approx 31 \text{ SVU}$$

Each firm expands its capacity to 31 vessels, compared to the optimum of 8 vessels.

Strategic Behavior by Firms

Because the number of firms is here assumed to be fixed, there is an opportunity for strategic behavior, in which individual firms limit their capacity without any specific agreement. We next discuss a simple game-theoretic model for this situation.

Consider the capacity decision of firm i, assuming that the total existing capacity of the other $N - 1$ firms equals Δ. If firm i invests in capacity E_i, then its share of the total catch Q will equal S_i, where

$$S_i = \frac{E_i}{E_i + \Delta} \qquad (4.25)$$

and the firm's net annual operating revenue is

$$R_i = pS_iQ - \frac{c}{qx}S_iQ = \left(p - \frac{c}{qx}\right)S_iQ \qquad (4.26)$$

(note that $H = qx(E_i + \Delta)\cdot D = Q$, which implies that $D = Q/qx(E_i + \Delta)$, so that the firm's variable cost equals $cE_iD = (c/qx)S_iQ$). The firm's present value is

$$PV_i = R_i\left(\frac{1+r}{r}\right) \qquad (4.27)$$

The firm chooses capacity E_i to

$$\text{maximize } PV_i - c_fE_i \qquad (4.28)$$

This can be written as

$$\text{maximize }\left(A\frac{E_i}{E_i + \Delta} - c_fE_i\right) \qquad (4.29)$$

where $A = (p - c/qx)(1 + r)Q/r = PV$. This implies that

Table 4.1. *Individual firm capacity under three investment scenarios.*

No. firms N	Firm's Capacity (SVU)			
	Optimal $\hat{E}/N$	Open access $E_{\max,\mathrm{OA}}$	Strategic E_i	Total Fleet (Strategic)NE_i
2	20	77	38.5	77
3	13.3	51.3	34.2	102.7
4	10	38.5	28.9	115.5
5	8	30.8	24.5	123.2
20	2	7.7	7.1	146.3

$$\frac{A\Delta}{(E_i + \Delta)^2} = c_f$$

If all firms adopt the same capacity E_i we have $\Delta = (N-1)E_i$ and the above equation reduces to

$$E_i = \frac{N-1}{N^2} \frac{A}{c_f} = \frac{N-1}{N^2} \frac{PV}{c_f} \qquad (4.30)$$

This should be compared with the pure open-access (non-strategic) capacity $E_{\max,\mathrm{OA}} = PV/Nc_f$ as in Eq. (4.24). We see that $E_i < E_{\max,\mathrm{OA}}$ with $E_i \approx E_{\max,\mathrm{OA}}$ if N is large. For the numerical example we obtain the results shown in Table 4.1. Total fleet capacity under strategic investment decisions is lower than under the assumption of pure open access (154 SVU), but still much higher than the optimum (40 SVU).

(The foregoing model does not apply when $N = 1$. In this case the firm's optimal capacity is the same as the overall optimum, $\hat{E} = 40$ SVU. The firm maximizes its present value, given the annual quota Q, by minimizing its costs. This implies that the firm's optimal capacity is the minimum capacity required to catch its quota, i.e. $\hat{E}$.)

Why is it that, with strategic individual investment behavior, total fleet capacity is predicted to expand far beyond the optimum, even if the total annual catch Q is fixed, and entry to the fishery is limited to N firms? At this stage in the book, the reader should be able to answer this question with complete assurance. The explanation is the same "tragedy of the commons" mechanism that has dominated earlier discussions in the book. Unless the firms can establish a binding agreement not to acquire excess capacity (or to overfish, but this possibility is here ruled out by the TAC assumption), game-theoretic aspects force the firms to overexpand. Any firm that fails to do so will catch less than its share of the TAC, usually much less.

The history of Antarctic whaling again provides an illustration (Clark and Lamberson 1982). In 1944 the newly established International Whaling Commission established an annual TAC of 16,000 blue whale units for combined-species whaling in the Antarctic. Six whaling nations were active in the Antarctic at that time, or soon after. Because this TAC was not allocated among the participating states or whaling companies, rapid expansion of the whaling fleets ensued. For example, in 1946–47 15 factories with 129 catchers took a total of 15,400 BWU in a season lasting 112 days, whereas in 1951–52 a fleet of 19 factories and 263 catchers took 16,000 BWU over a 64-day season (Clark and Lamberson 1982, p. 109).

Beginning in 1961 the whalers agreed to allocate the TAC among participating states. As a result, the 1962 whaling season lasted 115 days. As explained later, this outcome was probably close to the optimum, at least in terms of capturing the TAC.

Do fishermen really make strategic investment decisions? This idea is in line with our previous assumption of rational expectations. In practice, investment decisions clearly involve many uncertainties. Nevertheless it seems likely that fishermen do try to forecast future revenues, at least over several years, on the basis of whatever information they possess. But predicting how much fishing capacity will develop, and how rapidly, does seem to be a difficult task. We next consider this question for individual, licensed vessel owners.

Capital Stuffing: Nonlinear Costs

The foregoing model pertains to the capacity (SVU) decisions of N firms. Similar arguments apply to the capacity decisions of N independent vessel owners in a TAC-regulated fishery. We assume that the licenses restrict various features of the vessel, such as length, tonnage, HP and so on. Unless the restrictions control virtually every component of the vessel and its gear, however, the owner still has the option of increasing his fishing capacity E_i, but at a nonlinear capital cost $c_f(E_i)$, with

$$c'_f(E_i) > 0 \text{ and } c''_f(E_i) > 0$$

We use the strategic model, Eqs. (4.25)–(4.28). Equation (4.29) now becomes

$$\text{maximize} \left(A\frac{E_i}{E_i + \Delta} - c_f(E_i) \right) \qquad (4.31)$$

where $A = PV$, total fleet present value. By differentiation, followed by the substitution $\Delta = (N - 1)E_i$ this implies that

$$c'_f(E_i)E_i = \frac{N - 1}{N^2}PV \qquad (4.32)$$

(similar to the linear case, Eq. 4.30). For example, if $c_f(E_i) = c_1 E_i + c_2 E_i^2$ we have

$$c_1 E_i + 2c_2 E_i^2 = \frac{N - 1}{N^2}PV \qquad \text{nonlinear case}$$

$$c_1 E_i = \frac{N - 1}{N^2}PV \qquad \text{linear case}$$

In contrast to the linear case, marginal cost of capacity is increasing in the nonlinear model, and this implies a lower level of vessel capacity in the nonlinear case, other things being equal.

Once again let us use our numerical illustration of Sec. 3.4, for which $PV = \$77$ million. Previously we assumed a constant vessel cost of $c_f = \$500,000/\text{SVU}$. If we now suppose that $N = 40$ vessel licenses are issued, we get from Eq. (4.30), for the strategic model

$$E_i = 3.75 \text{ SVU}$$

Imagining that the 40 licensed vessels initially each possess one SVU of fishing capacity (recall that the optimal capacity in this model is 40 SVU), this prediction would indicate that vessel owners would attempt to more than triple vessel capacity.

Next, suppose that, foreseeing this possibility, the licensing agency imposes regulations designed to control vessel capacity. The regulations are partly successful, in that they change the capacity cost from $c_f E_i = \$500,000E_i$ to

$$c_f(E_i) = 200,000E_i + 300,000E_i^2$$

(this keeps the cost of one SVU at \$500,000). Equation (4.32) implies now that

$$E_i = 1.61 \text{ SVU}$$

i.e., vessel owners will increase the fishing capacity of their vessels by 61%. This results in a 61% increase of fleet capacity. What happens next depends on whether the licensing agency realizes what has happened. If so, they will shorten the fishing season by 38%; if not, the fish population will be overfished.

Though entirely artificial, our numerical example seems to reflect the reality of TAC, limited-access fisheries, which often experience significant expansions of fishing capacity as fishermen adopt technological improvements to their vessels and gear, a process referred to as capital stuffing. (Recall from our game-theoretic model in Sec. 3.4 that capital stuffing can be expected to occur even if it actually *reduces* the profitability of each vessel.)

4.3 Individual Fishing Quotas

The use of tradeable individual quotas has often been recommended for the management of open-access resources. To mention one example outside fisheries, tradeable pollution permits are now employed in many countries. Such permits, or quotas, are considered preferable to complex pollution regulations because they encourage firms to invent and use efficient low-pollution technology. Efficient use of information is also encouraged, as pollution-control decisions are made by individual firms, not by a central government agency. Of course the actual emissions must be rigorously monitored, and penalties imposed if the quota limit is exceeded. The total quota (i.e. sum of the individual quotas) is determined by the government on social cost–benefit considerations.

In a way, tradeable quotas are similar to taxes. To continue with the pollution example, imagine that a certain firm owns a factory that emits a pollutant at the rate of x tonnes/yr. To do so, it must hold a quota for x tonnes. Quota units can be bought or sold, on a quota market, at a price of \$$m$/tonne/yr. By reducing its emissions by Δx tonnes, the firm could gain $m\Delta x$ by selling off its unused quota. Conversely, additional quotas could be purchased, allowing the firm to increase its pollution level. This explains how pollution quotas are related to pollution taxes: causing pollution incurs an opportunity cost, proportional to the level of pollution. Note that this argument depends on the assumption that the quotas are tradeable.

Individual Fishing Quotas
The driving force behind both overfishing and overcapacity is the attempts of individual fishermen to maximize their economic benefits.

These attempts have similar implications in open-access, regulated open-access, and limited-entry, regulated open-access fisheries. Specifically, they result in overcapacity and the dissipation of profits. These in turn generate political pressure for persistently high annual quotas.

Individual fishing quotas (IFQs) have the potential to break this vicious cycle of regulation and counter-reaction. In the context of our present model, assume now that each vessel (or firm) is allocated an annual catch quota $Q_i = Q/N$. Catches will be carefully monitored; any fisherman apprehended illegally offloading fish in excess of his quota can expect severe penalties, such as loss of his license. The sum of the individual quotas equals the TAC. Also, only firms, or fishermen, that own quotas are allowed to fish; poachers are rigorously excluded.

We consider a model of an individual IFQ owner's daily decisions. The IFQ fisherman chooses his daily effort level E_i, and the number of days T_i per season to be spent fishing. We have

$$0 \leq E_i \leq E_{i,\max} \tag{4.33}$$

$$O \leq T_i \leq T_{\max} \tag{4.34}$$

where $T_{\max}$, the maximum season length, is determined by the managers, or possibly by the seasonal availability of the resource. Also $E_{i,\max}$ is the fisherman's maximum daily effort capacity, as determined by his particular configuration of vessel and gear.

For simplicity we ignore within-season variations in the stock biomass x. Thus all days within the season $0 \leq t \leq T_{\max}$ are the same, so that the fisherman's decision problem is independent of time. (Many of these simplifying assumptions could be replaced with more realistic ones, with possibly interesting results. For example, random daily variations in catch, or price, would be worth studying. Space limitations prevent a complete discussion of such details here.)

The fisherman's annual catch is

$$H_i = \begin{cases} qxE_iT_i & \text{if this is} \leq Q_i \\ Q_i & \text{otherwise} \end{cases} \tag{4.35}$$

His net daily revenue is

$$R_i = pqxE_i - c_i(E_i) \tag{4.36}$$

where $c_i(E_i)$ denotes daily effort cost. As elsewhere in this book, we suppose that

$$c_i'(E_i) > 0 \text{ and } c_i''(E_i) \geq 0 \tag{4.37}$$

(see Figure 2.15). As noted in Sec. 2.6, it is reasonable to suppose that daily effort costs increase nonlinearly as $E_i \to E_{i,\max}$. The special case of a constrained linear cost model,

$$c_i(E_i) = c_0 + c_i E_i \quad (0 \le E_i \le E_{i,\max})$$

is also worth considering; we treat this case separately.

Net annual revenue $R_{i,\mathrm{ann}}$ for the fisherman is given by

$$R_{i,\mathrm{ann}} = (pqxE_i - c_i(E_i))T_i \tag{4.38}$$

Thus the fisherman's decision problem can be written as

$$\underset{E_i, T_i}{\text{maximize}} \ (pqxE_i - c_i(E_i))T_i \tag{4.39}$$

$$\text{subject to } qxE_iT_i \le Q_i \tag{4.40}$$

$$\text{and } 0 \le E_i \le E_{i,max} \text{ and } 0 \le T_i \le T_{\max} \tag{4.41}$$

This problem is mathematically identical to the optimization model for a derby fishery as discussed in Sec. 2.6. Here, however, we are considering individual rather than fishery-wide optimization, with the added twist of IFQs. It is an important prediction that the use of IFQs can result in optimal fishing behavior. To emphasize this point we repeat the analysis here.

To begin with, we treat all model parameters as given constants, except for the individual's catch quota Q_i. Not surprisingly, the fisherman's income-maximizing strategy depends on how large his quota is—a small quota induces a different daily fishing strategy from that of a large quota. For simplicity we now omit the subscript i from our formulas. We consider two cases.

Case A: The annual quota is not taken: $H = qxET < Q$. In this case the fisherman fishes for the entire season ($T = T_{\max}$) but fails to catch the full quota. Equation (4.39) implies that daily effort E maximizes daily net revenue $pqxE - c(E)$. Letting E_0 denote this effort level, we see that

$$c'(E_0) = pqx \qquad \text{if } c'(E_{\max}) \le pqx \tag{4.42}$$

$$E_0 = E_{\max} \qquad \text{otherwise} \tag{4.43}$$

Thus in case (A) the fisherman simply maximizes his daily net income R. Note, by the way, that this is the same behavior that the fisherman would use in the TAC-regulated derby fishery (Sec. 2.6). The problem in that case is that overcapacity will tend to develop, resulting in a

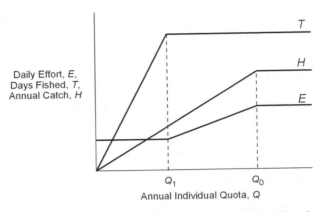

Figure 4.3 Individual daily effort E, number of days fished T, and annual catch H, in an IFQ fishery.

shortened fishing season. What transpires in the IFQ fishery, in terms of capacity, is discussed later.

We define the season's catch level Q_0 by

$$Q_0 = qxE_0T_{max} \tag{4.44}$$

Then the result for Case A is:

If $Q > Q_0$ then

$$E = E_0$$
$$T = T_{max}$$
$$H = qxE_0T_{max} < Q$$
$$R = (pqxE_0 - c(E_0))T_{max} \tag{4.45}$$

To repeat, these are optimal outcomes for the individual IFQ fisherman, with a large annual quota $Q > Q_0$—see Fig. 4.3.

Case B: The quota is taken: $H = qxET = Q$. In this case Eq. (4.39) becomes

$$\underset{E,T}{\text{maximize}} \left(p - \frac{c(E)}{qxE} \right) Q \tag{4.46}$$

subject to the condition $H = qxET = Q$. Thus the optimal daily effort E now minimizes average cost $c(E)/E$, subject to $qxET = Q$.

Define the effort level E_1 by

$$c'(E_1) = c(E_1)/E_1 \tag{4.47}$$

(thus E_1 minimizes $c(E)/E$ unrestrictedly), and also define Q_1 by

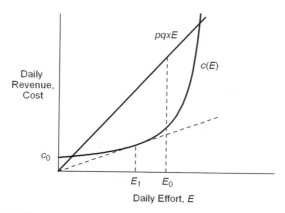

Figure 4.4 In an economically viable fishery we have $pqxE > c(E)$, which implies that $E_1 < E_0$.

$$Q_1 = qxE_1 T_{\max} \tag{4.48}$$

We have

$$0 \le E_1 < E_0 \tag{4.49}$$

(this is explained in Fig. 4.4). It is never optimal to use a positive daily effort level less than E_1, but the quota constraint implies that the optimum occurs at $E > E_1$ for a range of Q values. The result is:

If $Q_1 < Q < Q_0$ then

$$T = T_{\max}$$
$$H = qxET_{\max} = Q$$
$$E = Q/qxT_{\max} > E_1 \tag{4.50}$$

If $Q \le Q_1$ then

$$E = E_1$$
$$H = qxE_1 T = Q$$
$$T = Q/qxE_1 < T_{\max} \tag{4.51}$$

(Readers familiar with constrained optimization techniques will recognize the first case here, Eqs. (4.50), as a corner solution with both constraints binding, $T = T_{\max}$ and $H = Q$.) Figure 4.3 depicts the optimal solution for all cases.

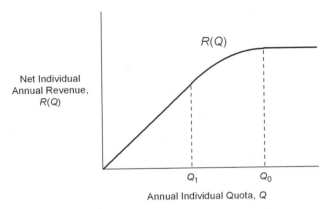

Figure 4.5 Net individual annual revenue (per vessel) $R(Q)$ as a function of the vessel's annual quota Q.

The expression for net annual revenue per vessel is obtained by substituting the above values into Eq. (4.38):

$$R_{\mathrm{ann}}(Q) = \begin{cases} [p - c(E_1)/qxE_1]Q & \text{if } Q \le Q_1 \\ [p - c(E_Q)/qxE_Q]Q & \text{if } Q_1 < Q < Q_o \\ [p - c(E_0)/qxE_0]Q_0 & \text{if } Q \ge Q_0 \end{cases} \qquad (4.52)$$

Here E_Q is defined by Eq. (4.50), i.e. $E_Q = Q/qxT_{\max}$. The graph of $R_{\mathrm{ann}}(Q)$ is shown in Fig. 4.5. Notice that $R_{\mathrm{ann}}(Q)$ is linear for $Q \le Q_1$, and constant for $Q \ge Q_0$.

As noted earlier, this individual optimization behavior, in an IFQ system, is *exactly the same* as the optimal fleet-wide fishing strategy in a TAC-regulated fishery—see Sec. 2.6 and Fig. 2.17. This encouraging prediction is central to the topic of individual quotas, or IFQs. (To see that the two results are the same, note that the TAC model can be transformed mathematically to the IFQ model, by writing $Q_i = Q/N$, where Q is the TAC and N the number of vessels. This change of variables in fact transforms Fig. 2.17 into Fig. 4.3.)

To reiterate, an IFQ system can in principle achieve optimal (profit maximizing) fishing patterns, without the use of additional complex regulations. For the moment, however, we must treat this as a model-specific result. Shortly, we will ask what happens if we make the model more realistic in various ways. For example, what if all vessels are not identical (as assumed in the model)? How do we determine the optimal allocation of quotas in this case? What is the optimal fleet size? Will IFQs automatically induce a reduction in fleet size, in the case that existing capacity is excessive? We address such questions after some further discussion of our present model predictions.

Characteristics of the IFQ System

What is it that distinguishes the two "optimal" effort levels E_1 and E_0? (We examine a numerical example, below.) Recall that E_0 maximizes the fisherman's net daily income $R = pqxE - c(E)$, whereas E_1 minimizes his average daily cost $c(E)/E$. Intuitively, both would seem to be desirable strategies—why this difference?

The explanation is that E_0 is the fisherman's optimal strategy if the season length T is specified, but not his annual catch. On the other hand E_1 is the fisherman's optimum if his annual catch is specified, but T is not. In the latter case, for example, the fisherman's gross annual revenue $pH = pqxE$ is specified, so his only concern is to capture his quota at the lowest total cost, which implies minimizing average daily cost $c(E)/E$.

Also, E_1 is optimal for the fishery as a whole, for the same reason, namely that the TAC is specified. (Admittedly, the managers may attempt to achieve the target TAC by specifying a season length, but it is the catch that they are primarily concerned with.)

I dwell on this point for two reasons. First, and specifically, it is important to understand why IFQs are considered desirable (we will have much more to say on the pros and cons of IFQs). Second, it is a fundamental principle underlying this book's modeling philosophy that the modeler needs to understand (and explain) both the assumptions built into his or her model, and also the reason for the model predictions. It is *not* acceptable for the modeler, when asked to explain his predictions, to say "Oh, it just comes out from the model." Mathematicians, economists, and computer modelers are often guilty of this attitude, which perhaps helps to explain the widespread distrust of models. Certainly the basic bioeconomic models used in fisheries have been drastically and persistently misunderstood.

For a numerical illustration we use the following values:

$$x = 10,000 \text{ tonnes}$$
$$q = 5 \times 10^{-5}/\text{SVU day}$$
$$Q = 80 \text{ tonnes (individual annual quota)}$$
$$T_{\max} = 200 \text{ days}$$
$$p = \$1,000/\text{tonne}$$
$$c(E) = 50 + 50E + 50E^2 \ \$/\text{day}$$

$$E_{\text{max}} = 3.0 \text{ SVU}$$

$$N = 25 \text{ vessels}$$

The values for x and q imply a daily catch rate $H = qxE = 0.5E$ (tonnes/day). First we consider the TAC-regulated, no IFQ case. The fisherman uses $E = E_0$ which maximizes $R = pqxE - c(E)$. Setting $c'(E) = pqx$ gives $E_0 = 4.5$, but since $E_{\text{max}} = 3.0$ we set $E_0 = 3.0$ SVU. The daily catch rate is then $qxE_0 = 1.5$ tonnes/day. If we suppose that the TAC fishery will be closed when the average vessel has caught 80 tonnes, then we have $T = 53.3$ days. The fisherman's net annual income is

$$R_{\text{TAC}} = (pqxE_0 - c(E_0))T = \$45,305$$

In the IFQ case we have $E = E_1$ where $c'(E_1) = c(E_1)/E_1$, so that $E_1 = 1.0$ SVU. Thus $T = 160$ days (i.e., IFQs extend the fishing season three-fold), and we have

$$R_{\text{IFQ}} = (pqxE_1 - c(E_1))T = \$56,000$$

The IFQ system increases the average income of fishermen by 23.6%. (We have $Q_1 = 100$ tonnes and $Q_0 = 300$ tonnes, so $E = E_1$ is the IFQ solution for the case that $Q = 80$ tonnes).

Next, what if $c_0 = 0$? (Recall that c_0 is the daily mobilization cost. This might be positive if the vessel returns to port each night, but not otherwise.) Now we obtain $E_1 = 0$ (Fig. 4.4), so that $Q_1 = 0$ also. The case $Q < Q_1$ thus does not arise. For the numerical example, we now have $c(E) = 50E + 50E^2$. Therefore

$$R_{\text{TAC}} = \$47,970$$

whereas the IFQ solution has $E = Q/qxT_{\text{max}} = 0.8$ SVU, and

$$R_{\text{IFQ}} = \$65,600$$

In this case the IFQ revenue is 36.7% larger than the TAC revenue.

Another special case, with different interpretations, occurs if the daily effort–cost function $c(E)$ is assumed to be linear: $c(E) = c_0 + c_1 E$. From Eq. (4.39) it now follows that $E_1 = E_0 = E_{\text{max}}$, i.e., optimal fishing always uses $E = E_{\text{max}}$. In other words, a fisherman's optimal behavior during the fishing season is the same under the derby fishery and the IFQ system. Also, his net annual income will be the same under both systems. The IFQs have no effect on fishermen's behavior.

Does this mean that IFQs are only worthwhile in situations where daily effort costs are nonlinear? Not at all. We will next consider the possibility of pooling quotas, in which case further economic benefits can occur. However, it should again be emphasized that nonlinear effort cost is probably the most realistic assumption. The derby fishery (Sec. 2.6) forces individual fishermen to operate at high effort intensity while the season remains open, which can reduce economic returns, as well as incurring other indirect costs, such as increased risk of accidents.

Pooling of Quotas

Assume now that catch quotas are allocated to individual vessels, but that quotas can be freely pooled, at least within any year. Such quota pooling could have several advantages. For example, the introduction of allocated quotas into a fishery with extreme excess capacity may result in a large number of small individual quotas. Efficiency might then be increased by pooling quotas and operating only a fraction of the initial fleet. The following model explains the conditions in which quota pooling can be beneficial.

We now let $c_A(Q)$ denote the total annual cost of harvesting an individual quota Q under an IFQ system. As before we assume that

$$c_A'(Q) > 0, \quad c_A''(Q) \geq 0 \quad \text{for } 0 \leq Q \leq Q_{\max} \tag{4.53}$$

We call $c_s = c_A(0)$ the annual setup cost; this would include all costs of preparing the ship for the year's season and sailing it to the fishing grounds. The variable annual cost $(c_A(Q) - c_S)$ can be calculated for the above IFQ model. Equations (4.50) and (4.51) imply that

$$c_A(Q) - c_S = \begin{cases} Qc(E_1)/qxE_1 & \text{if } Q \leq Q_1 \\ T_{\max}c(Q/qxT_{\max}) & \text{if } Q_1 \leq Q \leq Q_0 \end{cases} \tag{4.54}$$

This is a differentiable function, satisfying Eq. (4.53). The marginal cost function $c_A'(Q)$ is therefore given by

$$c_A'(Q) = \begin{cases} \gamma = \text{constant} & \text{if } Q < Q_1 \\ c'(E_A)/qx & \text{if } Q_1 \leq Q \leq Q_0 \end{cases} \tag{4.55}$$

where

$$\gamma = c(E_1)/qxE_1$$
$$E_A = Q/qxT_{\max}$$

This marginal cost function is shown in Fig. 4.6.

We first consider two quota holders, holding equal quotas Q_i and having identical cost functions $c_A(Q)$. If both vessels catch their original

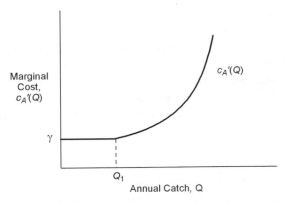

Figure 4.6 Marginal cost function $c'_A(Q)$ for annual catch Q.

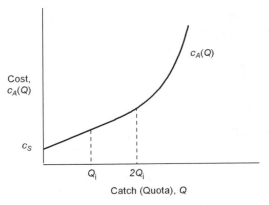

Figure 4.7 Cost function $c_A(Q)$ for catching the annual quota Q. Here c_S is the annual setup cost.

quota Q_i, the total cost of fishing is $2c_A(Q_i)$, whereas if the pooled quota $2Q_i$ is caught by a single vessel, fishing costs become $c_A(2Q_i)$. For the case shown in Fig. 4.7 we have

$$c_A(2Q_i) < 2c_A(Q_i)$$

so that pooling of the two quotas is more efficient than catching them with two vessels. (Note that $c_A(2Q_i) < 2c_A(Q_i)$ if Q_i is sufficiently small, but that this inequality is reversed if Q_i is large.) This raises the question of the optimal number of quotas to be pooled. It turns out that the optimal pooling strategy minimizes average cost per active vessel. Box 4.1 spells out the details.

Box 4.1. Optimal pooling of quotas.

Suppose $N >> 1$ vessels have the same individual quotas Q_i, and identical cost functions $c_A(Q)$ for harvesting the quantity Q. Suppose that these quotas are pooled into groups of N^* individual quotas, so that N/N^* vessels each catch $Q^* = N^*Q_i$. The total cost is then

$$\frac{N}{N^*}c_A(Q^*) = NQ_i\frac{c_A(Q^*)}{Q^*}$$

Thus total fishing cost is minimized if the pooled quotas Q^* minimize the average cost $c_A(Q^*)/Q^*$. Figure 4.8 shows a graphical technique for determining the optimal pooled quota size Q^*.

Figure 4.8 shows two possible cost function types, (a) nonlinear and (b) linear, both of which are encompassed by our theory. In the linear case the optimal pooled quota Q^* equals the vessel capacity Q_{max}. In the nonlinear case Q^* is less than full capacity, because costs increase sharply as Q approaches Q_{max}.

To illustrate, we again use our numerical example (Sec. 3.4) with

$$Q_{total} = 10,000 \text{ tonnes}$$
$$x = 50,000 \text{ tonnes}$$
$$q = .01/\text{SVU yr}$$
$$D_1 = .5 \text{ yr}$$
$$p = \$1,000/\text{tonne}$$
$$c = \$150,000/\text{SVU yr}$$
$$c_f = \$500,000/\text{SVU}$$

Effort costs are assumed to be linear, $c(E) = cE$, where E now denotes annual effort. However, there is now an annual setup cost

$$c_S = \$50,000/\text{SVU}$$

The optimal fleet size is the smallest fleet needed to catch the annual quota, $\hat{K} = Q/qxD_1 = 40$ SVU.

We can now compare the model predictions for three management scenarios based on this model: TAC-regulated open access, IFQs without pooling, and IFQs with pooling. First, for the open-access case our model now includes an additional annual

---

Box 4.1 *continued*

setup cost $c_S = \$50,000$/SVU. Open-access fleet capacity K_{OA} is therefore given by

$$PV - c_S K_{\mathrm{OA}}(1 + r)/r = c_f K_{\mathrm{OA}}$$

From Sec. 3.4 we have $PV = \$77$ million and $r = .10$/yr. Therefore we now have

$$K_{\mathrm{OA}} = 73 \text{ SVU}$$

The net present value of economic profits, including setup and capital costs, is zero by assumption. Annual vessel net operating revenues $R/K_{\mathrm{OA}} - c_S$ are positive however, and are equal to $45,890 /SVU.

Next, suppose that an IFQ system is introduced, by assigning annual catch quotas of $Q/K_{\mathrm{OA}} = 137$ tonnes/yr to each existing vessel. Because we are here assuming linear effort costs $c(E) = c \cdot E$, our IFQ model predicts no change in the fishermen's behavior, or in the economic outcome. (There would only be changes in the case that the open-access fishery were a derby fishery with nonlinear costs $c(E)$.)

If, however, pooling of quotas is permitted, our current model predicts that $N^* = 40$ vessels will actively fish, each taking its maximum catch of 250 tonnes/yr. This results in a total annual saving of $(73 - 40)c_S = \$1.65$ million, or \$22,603 per existing vessel. Thus the annual net revenues are increased from \$45,890 to \$68,493 for each of the original 73 vessels in the fishery. This equals the maximum economic rent for our model fishery.

Notice that optimal quota pooling implies optimal active fleet capacity. This is a principal feature of IFQs. But will optimal pooling automatically occur in an IFQ fishery, or do the managers need to impose pooling? Assuming that the owners of pooled quotas will share equally in the net revenues (whether they actually participate in fishing, or not) there will be an economic incentive to form quota pools of approximately optimal size. Each owner will earn a higher income by pooling than by catching his own quota. Likewise, unduly large pools will reduce profits for each owner. IFQs remove the incentive for capital stuffing, and re-

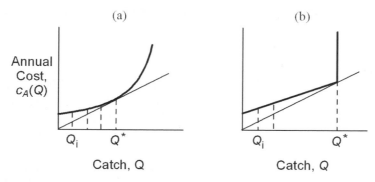

Figure 4.8 Graphical determination of the optimal pooled quota size Q^*. (a) nonlinear costs; (b) linear costs. See Box 4.1.

place it by an incentive for cost minimization. Pooling could be arranged by the fishermen themselves, without government involvement. However, government assistance in setting up pooling arrangements might be worthwhile, as occurred recently in the British Columbia roe-herring fishery.

So far, we have discerned two important advantages of IFQs. First, in a derby fishery (Sec. 2.6), IFQs can alter daily fishing patterns, which reduces overall costs, thereby preserving economic profits. In addition, by means of quota pooling, an IFQ system can also alter annual fishing patterns, reducing the number of actively participating vessels, thereby reducing total annual setup costs and preserving additional profits. In other words, IFQs directly counteract the "tragedy" of the common-pool, or open-access fishery, and induce economically efficient behavior on the part of the fishermen. This occurs simply because individual fishermen no longer find it necessary to compete for their share of the common resource pool.

Without specifically modeling the question, we can also conclude that IFQs will eliminate the motive for capital stuffing, which would be of no benefit to a fisherman whose annual quota is fixed. This does not mean, however, that IFQ fishermen will never wish to improve the efficiency of their vessels, as for example when new technology becomes available. If the additional capital expense will pay off by sufficiently reducing the costs of catching the annual quota, it will be undertaken. Such behavior is entirely rational, both from the individual's viewpoint and that of the economy as a whole.

But this raises a new question: would a given fisherman perhaps wish to increase (or decrease) his quota holdings? We next study this important, and controversial, issue.

Transferable Quotas

Unrestricted transferability of IFQs (which are then called ITQs) is desirable for reasons discussed below. But transferability may also be considered socially undesirable, because of the likelihood that quota ownership may become concentrated in a few individuals or fishing companies.

To analyze these questions, we first model the demand for quotas, as a function of the quota price m. As usual, the price of quotas will be determined by balancing supply and demand. In the case of a TAC-regulated fishery, supply is fixed at the annual TAC Q. (For now we continue to ignore fluctuations in the annual TAC.)

In the following analysis we finally drop the simplifying assumption that all fishermen (and their vessels) are identical in terms of cost functions. We suppose that vessel owner i holds a permanent ITQ allowing him to catch and land Q_i tonnes of fish per year. His annual net revenue is assumed to be

$$R_i(Q_i) = pQ_i - c_i(Q_i) \tag{4.56}$$

(see Eq. 4.54) and the present value of this revenue is

$$PV_i(Q_i) = \frac{1}{r_i} R_i(Q_i) \tag{4.57}$$

All symbols here have the same interpretation as before. The cost functions $c_i(Q_i)$ and discount rates r_i are owner-specific. We now use the expression R_i/r_i for the present value of a fixed annuity R_i, rather than $R_i(1 + r_i)/r_i$ as used previously. This amounts to ignoring the current year's payment. The new expression simplifies later formulas.

A quota market for purchase and sale of (permanent) quota units is assumed to exist. Let m denote the unit price of quota units. A given vessel owner i will then wish to purchase a marginal additional quota ΔQ if and only if this benefits him more than the cost, or in other words, if and only if

$$PV_i(Q_i + \Delta Q) - PV_i(Q_i) > m\Delta Q$$

Assuming that ΔQ is small, this becomes

$$\text{buy more quota if and only if } PV_i'(Q_i) > m \tag{4.58}$$

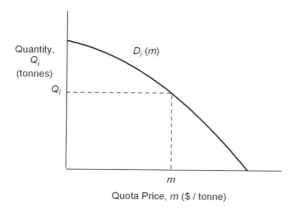

Figure 4.9 Vessel owner i's demand function for quotas.

where $PV_i'(Q_i)$ denotes the derivative of $PV_i(Q_i)$. This determines owner i's demand function $D_i(m)$ for quota units:

$$D_i(m) = Q_i \quad \text{where} \quad PV_i'(Q_i) = m \qquad (4.59)$$

—see Fig. 4.9. In other words, D_i is the inverse function of PV_i'. We now assume nonlinear costs, $c_i''(Q_i) > 0$. (The linear case is easily dealt with, but we skip the details). Then $D_i(m)$ is deceasing in m, as shown in Fig. 4.9. A higher quota price means that the owner wants to hold onto fewer quota units. Rewriting Eq. (4.58) gives the rule:

$$\text{buy more quota units if and only if } Q_i < D_i(m) \qquad (4.60)$$

where Q_i denotes the owner's present holding. Likewise, the owner will wish to sell some quota units if $Q_i > D_i(m)$. Consequently, owner i will buy or sell quota units until his final holding Q_i satisfies

$$PV_i'(Q_i) = m \qquad (4.61)$$

Note that this equation implies that the owner decides to retain the quota Q_i that maximizes his net present value

$$NPV_i(Q_i) = PV_i(Q_i) - mQ_i = \frac{1}{r_i}((p - m_i')Q_i - c_i(Q_i)) \qquad (4.62)$$

where $m_i' = r_i m$.

The industry demand function is the sum of individual demand functions:

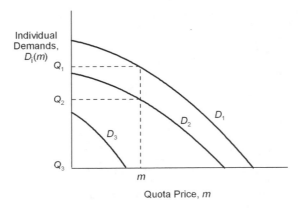

Figure 4.10 Three typical demand functions $D_i(m)$. Owner i retains quota $Q_i = D_i(m)$. Here, owner 3 sells his entire quota: $Q_3 = 0$.

$$D_{\text{industry}}(m) = \sum_i D_i(m) \qquad (4.63)$$

This also is a decreasing function of m. As is usual in economics, we assume that this demand function exists, even if it is never actually specified, or measured. The market-clearing condition is that supply equals demand, or

$$Q = D_{\text{industry}}(m) \qquad (4.64)$$

where Q is the total annual quota, or TAC. Thus, if Q is given, the quota price m is determined by Eq. (4.64), and the number Q_i of quota units retained by each vessel owner is then determined by Eq. (4.61).

Figure 4.10 shows three typical individual quota demand curves $D_i(m)$. The vertical line is the quota price m. Owners with higher demand curves hold more quota units than other owners. Some owners may even opt to sell their entire initial quotas, and leave the fishery. Under what conditions would this occur? High demand $D_i(m)$ corresponds to high marginal present value $PV_i'(Q_i)$, where

$$PV_i'(Q_i) = \frac{1}{r_i} \left(p - c_i'(Q_i) \right) \qquad (4.65)$$

Thus a high demand function can result from low marginal costs, or from a low discount rate. We conclude that

1. High marginal cost owners will tend to sell their quotas to lower marginal cost owners.
2. Owners with high time-preference (i.e. discount) rates will tend to sell their quotas to individuals with lower discount rates.

Are these features of an ITQ system desirable, or not? Also, is the final distribution of quota units Q_i optimal in some sense?

First we consider the optimality question. Let r_s denote the social rate of discount. The optimal distribution of quotas would then

$$\text{maximize } \sum_i PV_i^*(Q_i)$$

$$\text{subject to } \sum_i Q_i = Q$$

where

$$PV_i^*(Q_i) = \frac{1}{r_s}(pQ_i - c_i(Q_i))$$

The corresponding Lagrangian is

$$\mathcal{L} = \sum_i PV_i^*(Q_i) - \lambda \sum_i Q_i$$

and the necessary conditions for a maximum are

$$\frac{\partial \mathcal{L}}{\partial Q_j} = PV_j^{*'}(Q_j) - \lambda = 0 \quad j = 1, 2, \ldots, N$$

These are the same as Eq. (4.61), with $\lambda = m$, except that now all the discount rates r_i are replaced by r_s. We see that if all vessel owners adopt the social discount rate, then quota transfers will achieve the optimal result. Conversely, if some owners adopt different rates from other owners, the resulting distribution of quotas will be suboptimal.

A simple example will illustrate how this could work out in practice. Suppose that all owners have identical cost functions $c(Q_i)$, and that all but one owner uses the social discount rate r_s. The remaining owner has $r_1 > r_s$. If m is the quota price, we then have from Eqs. (4.61) and (4.65)

$$c'(Q_1) = p - r_1 m \text{ and } c'(Q_i) = p - r_s m \quad (i > 1)$$

Hence $c'(Q_1) < c'(Q_i)$, so that $Q_1 < Q_i$ ($i \neq 1$). The higher-discount-rate owner retains a smaller quota than other owners. The reason is straightforward; the owner with a high personal discount rate prefers to sell off more of his quota than other owners. It is even possible that this owner would sell off his entire original quota—"take the money and run," in other words.

Quota Leasing

In addition to permanent quota transfers, short-term leasing of quotas often occurs. Very short-term quota transfers may be a response to random annual fluctuations in the annual catches of different vessels. Vessel owners who have caught their annual quotas early may wish to obtain additional quota units from owners who do not expect to catch their full quota.

Annual quota leasing may also occur, if some quota owners prefer to lease some or all of their quotas on a seasonal basis. The question is, why would a quota owner favor leasing all or part of his quota rather than using it himself, or selling it? Assuming that markets exist for both quota transfers and leases, how would the prices be related? Without developing an explicit two-market model of quota transfers, all we can say is that we expect to find that

$$r_1 m_{\text{transfer}} < m_{\text{lease}} < r_2 m_{\text{transfer}} \tag{4.66}$$

where r_1, r_2 are the discount rates of quota leasers and renters, respectively, and the ms are the indicated prices.

Would non-fishermen ever be motivated to buy up quotas and then lease them back to active fishermen? And would some fishermen be willing to sell their quotas and then rent them back again on an annual basis? The answer to both questions is yes, provided that $r_1 < r_2$ (with a sufficient difference to cover transaction costs). Someone with a sufficiently low discount rate might even buy up all the quotas and then lease them back to the fishermen.

This is exactly what has happened in several ITQ fisheries. For example, the Vancouver Sun of March 16, 2004, reported that many fishermen participating in the British Columbia roe-herring fishery had rented quota licenses for the brief annual herring fishing season. The vice president of the United Fishermen and Allied Workers Union was quoted as saying that "Unfortunately the renting of licenses is all about greed by the guys who own the quotas—dentists in Toronto or businessmen in South Korea—and the guys who have to buy them to make a living." He also said that fishermen pay about $15,000 Canadian to "rent" a license allowing them to catch 10 tonnes of roe herring during the season.

Finite-Term Quotas

For future reference, we next consider the case in which individual quotas have a finite term of Y years. The present value expression now becomes

$$PV_{i,Y}(Q_i) = (pQ_i - c_i(Q_i)) \left(\frac{1}{1 + r_i} + \cdots + \frac{1}{(1 + r_i)^Y} \right)$$
$$= (pQ_i - c_i(Q_i))D_{i,Y} \tag{4.67}$$

where $D_{i,Y}$ is shorthand for the sum above:

$$D_{i,Y} = \frac{1}{r_i} \left(1 - \frac{1}{(1 + r_i)^Y} \right) \tag{4.68}$$

Note that if $Y = \infty$ we have $D_{i,Y} = 1/r_i$ as in Eq. (4.57).

The equations that now determine the market price m_Y for quotas, and the distribution of quotas Q_i, are modifications of the previous equations. Specifically

$$PV'_{i,Y}(Q_i) = m_Y \tag{4.69}$$

$$\sum_{i=1}^{N} Q_i = Q \tag{4.70}$$

Since the present value $PV_{i,Y}(Q_i)$ is an increasing function of the term Y, we conclude that the quota price m_Y is also increasing in Y. Longer-term ITQs are worth more than short-term ITQs. Short-term ITQs acknowledge that the state retains ownership of the resource, whereas indefinite ITQs amount to a transfer of resource ownership, unless the ITQs include payment of user royalties. But very short-term (e.g. annual) ITQs may involve large uncertainties for the fishermen. Any incentive for resource conservation may disappear as the terminal date of a fisherman's quota approaches.

To summarize, we have now identified a third advantage of IFQs, provided that they are transferable (i.e., ITQs). Namely, ITQs can motivate the optimal redistribution of quotas among fishermen. However, this advantage is accompanied by a potentially serious drawback, to which we have already alluded. We next consider this drawback in greater detail.

The IFQ Controversy

The use of IFQs (and especially ITQs) has been controversial in the USA and elsewhere. For example, in the Sustainable Fisheries Act of 1996 (an amendment of the Magnuson-Stevens Fishery Conservation and Management Act of 1976), "Congress placed a moratorium on the ability of the regional fishery management councils to develop or submit any fishery management plan using IFQs until October 1, 2000." (Ocean Sciences Board 1999, p. 14). As this report states, IFQs have evoked considerable controversy "because of [their] potential for creating windfall benefits to the initial recipients, the privileges that IFQs create, and the potential for decreasing employment and changing social and economic relationships among individuals and communities" (*ibid*, p. 14).

These statements clearly reveal a major problem with IFQs and ITQs. Let us summarize the situation.

Fishery management traditionally ignored the entire question of resource economics, concentrating instead on controlling fishing activities so as to prevent overfishing and stock depletion. Where this has succeeded it has almost invariably led to severe overcapacity of fishing fleets. Overcapacity destroys profitability, and may also increase the difficulties of fishery management, since sustainable TACs are often insufficient to support a profitable fishery in the presence of severe overcapacity.

Individual fishing quotas have the potential to resolve both the overfishing and overcapacity problems, while also preserving profitability. However, awarding IFQs free of charge to special privileged individuals amounts to a give-away of valuable public assets. This, no doubt, is why IFQs are controversial, and properly so, and this is probably also the reason that the discussion has bogged down. The fishing industry, accustomed to free access (in the financial sense) to the resource, in many cases accompanied by generous subsidies, naturally opposes any introduction of resource royalties or fees. Also, the fishing industry almost always exerts a strong influence over management decisions, with agreements often being reached behind closed doors with little or no public input or knowledge. The public interest is often simply not represented at the management level. Yet the public remains the ultimate owner of most marine resources, at least those that are located within 200-mile coastal zones.

Failure to levy substantial rent-capturing royalties in an IFQ or ITQ fishery has many side effects, including:

1. The initial allocation of quotas becomes difficult and controversial. Winners (quota recipients) may be greatly enriched.
2. Transferability of quotas may be opposed, on the grounds that this would allow quota recipients to enjoy a windfall gain without doing any actual work.

3. If quota transfers are in fact prohibited, fishermen having a legitimate desire to leave the fishery will find it difficult to do so, given that their quotas will expire if not used. Also, consolidation of small quotas into larger, more efficient units will be prevented (although pooling on a short-term basis might be allowed).
4. If transfers are permitted, quota ownership may rapidly become concentrated in the hands of a few wealthy entrepreneurs. Regulations that prohibit the acquisition of quota holdings by non-fishermen may be difficult to enforce.
5. If quota ownership is transferred to non-fishermen, many active fishermen will be forced to rent quotas on an annual basis. Resource profits will then mostly accrue to the quota owners.
6. Fishermen who rent quotas will have little incentive to practice or support resource conservation measures.
7. The anticipation of future royalty-free IFQ or ITQ systems will entice additional overcapacity, resulting in the speculative dissipation of the profits that the system is intended to preserve.

Item (7), for which I acknowledge the contribution of Gordon R. Munro, has not to my knowledge previously been considered in the literature. As we have seen elsewhere, anticipation by fishermen of a future management strategy that will enhance the profitability of the fishery, almost inevitably precipitates a round of capacity expansion. If the anticipated changes turn out to be realized, the expansion will exactly dissipate the intended gains in future profits. This difficulty would disappear if fishermen knew in advance that future windfall gains would be extracted by means of royalty charges.

Until the question of resource user fees is openly addressed and resolved in an economically effective and socially equitable manner, rational management of marine fisheries will remain unlikely. It is not as if user fees in publicly owned resources, such as off-shore oil and minerals, or national forests, are anything unusual. Surely the next logical step after establishing extended fisheries jurisdiction, is the recognition that coastal states have the obligation to extract significant revenues from their fishery resources. Section 4.5 discusses this question in greater detail.

Combining IFQs and Catch Royalties
We wish to study the effect of a catch royalty on the IFQ fishery, and

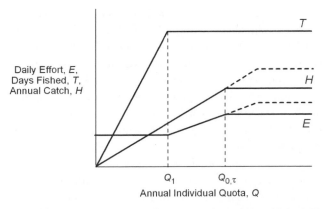

Figure 4.11 Individual daily effort E, number of days fished T, and annual catch H in an IFQ fishery with catch royalties. (The dotted lines show the result without catch royalties, as in Fig. 4.3.)

on the ITQ fishery. For the first case we use our IFQ model, replacing Eq. (4.39) by

$$\underset{E_i,T_i}{\text{maximize}} \; [(p-\tau)qxE_i - c_i(E_i)]T_i \qquad (4.71)$$

subject to the constraints in Eqs. (4.40)–(4.41). This problem is mathematically the same as before, with p replaced by $p-\tau$. By checking through Eqs. (4.42)–(4.51), we see that the only effect of the royalty τ is to replace the total quota threshold Q_0 by $Q_{0,\tau}$, where

$$Q_{0,\tau} = qxE_{0,\tau}T_{\max} \text{ where } c'(E_{0,\tau}) = (p-\tau)qx \qquad (4.72)$$

(or else $E_{0,\tau} = E_{\max}$ as in Eq. (4.43)). The other threshold Q_1 is unchanged. Figure 4.11 shows the result; this figure can be obtained by starting with Fig. 4.3, the shifting $Q_{0,\tau}$ to the left and lowering the curves for H and E for $Q > Q_{0,\tau}$ as shown.

What does Fig. 4.11 tell us about the effect of combining IFQs and catch royalties? First, if $Q < Q_{0,\tau}$ there is no effect on E, T or H. Otherwise E and H are reduced by the royalty. This occurs because the fisherman maximizes his daily net revenue when Q is large (and not taken). In other words, the catch royalty does not affect fishing behavior unless the fisherman has excess quota that he does not use, i.e., $Q > Q_{0,\tau}$. However, the royalty does have the effect of decreasing the threshold $Q_{0,\tau}$. If the royalty is initiated at the same time as the

IFQ system itself, the result could be a reduction in political pressure for a large TAC.

Of course the fisherman's annual revenues are reduced by the catch royalty. Net revenues R_τ are given by Eq. (4.52), with p replaced by $p - \tau$ and Q_0 by $Q_{0,\tau}$.

Sufficiently large royalties can shut down the fishery. Define τ_1 by

$$(p - \tau_1)qx = c'(E_1) \tag{4.73}$$

Then if $\tau > \tau_1$ fishing is unprofitable at any effort level—see Fig. 4.4. In this case Fig. 4.11 collapses, with $E = H = T = 0$.

Conservation Incentives

Under what circumstances can it be expected that fishermen will support conservationist management strategies? Here we assume that greater conservation means that a higher stock level x is maintained. In the case of a depleted population, greater conservation thus requires a reduction in current TACs to allow for stock rehabilitation. Once the rehabilitation is complete, larger TACs become possible. Annual revenues are increased, both because TACs are larger and because fishing costs are smaller than previously. In addition, the risk of population collapse is reduced.

Fishermen do not always favor such rehabilitation programs, however. In part this may be due to future discounting (Chapter 2), but another important consideration is the possibility that the existing fishermen may never actually reap the ultimate benefits of conservation. In a regulated open-access fishery, for example, successful stock rehabilitation will usually attract new entrants, with the result that profits will continue to be dissipated. One potential advantage of IFQs (more precisely, IFQ shares) is that quota owners can expect to gain from resource conservation—assuming of course that the government does not respond to increased catches by issuing additional quotas. Resource conservation will thus be recognized as a bona fide financial investment. Experience with several current IFQ systems supports this assertion—see Chapter 6.

What is the effect of catch royalties on conservation incentives? If the royalty captures most of the profits from fishing, and if the royalty rate is increased whenever conservation succeeds, then much of the fishermen's incentives favoring conservation will be eliminated. But if a good portion of the increased revenues from conservation are allocated to IFQ owners, the conservation incentives will be retained. Perhaps a 50-50 division of increased revenues between fishermen and resource owners (the state) would be a fair compromise. (Determining what royalty rate τ achieves a 50-50 split may not be easy, especially given the widespread tendency of fishermen to conceal their financial results. It seems to me that the resource owner has every right to access these financial results.)

But there still remains the dilemma of anticipation. To repeat, once an IFQ program is under discussion, potential fishermen will be motivated to enter the fishery (and existing fishermen to increase their fishing capacity) in order to be eligible for future quotas. To thwart this behavior it will be necessary to announce the government's intention to charge catch royalties on future catches.

Thus royalties (or user fees) are not merely a question of social equity in the distribution of resource revenues, but are also necessary to control the anticipatory dissipation of future revenues. Although the suggested 50-50 split would reduce dissipation to some extent, some would still occur. There may be no perfect resolution of this dilemma.

Catch Royalties in an ITQ Fishery

We next consider the effect of a catch royalty on an ITQ fishery. The analysis differs from the IFQ case because we now need to include the effect of the royalty on the quota price. Equations (4.56) and (4.57) now become

$$PV_{i,\tau}(Q_i) = \frac{1}{r_i}[(p - \tau)Q_i - c_i(Q_i)] \tag{4.74}$$

The demand equation of Eq. (4.61) is now $PV'_{i,\tau}(Q_{i,\tau}) = m_\tau$, or

$$c'_i(Q_{i,\tau}) = (p - \tau) - r_i m_\tau \tag{4.75}$$

where m_τ is the quota price with catch royalty τ.

Suppose first that all vessel owners have the same cost function $c(Q)$ and discount rate r. Then

$$c'(Q_{i,\tau}) = (p - \tau) - rm_\tau$$

Here all quotas $Q_{i,\tau}$ are equal. Since the TAC $Q = \Sigma Q_{i,\tau}$ is the same with or without royalties, we conclude that $Q_{i,\tau} = Q_{i,0}$ where $Q_{i,0}$ are the individual quotas with zero royalty, so that

$$c'(Q_{i,0}) = p - rm$$

Therefore we have $p - rm = p - \tau - rm_\tau$, or

$$m_\tau = m - \tau/r \tag{4.76}$$

The catch royalty τ (paid every year) has present value τ/r, which is the amount by which the price (or value) of a quota unit is reduced.

Next, suppose that there are two groups of fishermen and vessels, having the same cost functions, but different discount rates r_1, r_2 with

$$r_1 < r_2 \tag{4.77}$$

The quota demand functions are then determined by

$$c'(Q_{j,\tau}) = p - \tau - r_j m_\tau \tag{4.78}$$

We therefore have

$$\tau + r_j m_\tau > r_j m \Leftrightarrow Q_{j,\tau} < Q_{j,0} \tag{4.79}$$

Now, if one group increases its quota holdings under the royalty, the other must decrease its holdings (because $\Sigma Q_i = Q$ always). Since $m_\tau < m$, we must therefore have

$$\frac{\tau}{r_2} < m - m_\tau < \frac{\tau}{r_1} \tag{4.80}$$

by Eq. (4.77). Thus Eq. (4.79) implies that group 1 (low discount rate) reduces its quotas whereas group 2 (high discount rate) increases its quotas, relative to the case of zero royalties.

We also know from our earlier discussion that low-discount-rate fishermen hold higher quotas than high-rate ones, *ceteris paribus*, and this remains true with royalty payments. What happens, therefore, is that royalties tend to even out the quota holdings of fishermen with different discount rates. It seems likely that this increases the social welfare function for the fishery as a whole, but I will not try to prove this here.

Other Types of Royalty

The preceding discussion assumes that royalties are charged on actual landings. Alternatives include annual charges based on individuals' quota holdings, or net annual revenues. Dealing first with quota-based royalties, we note that the annual catch will usually be close to the quota. Thus quota-based royalties are basically the same as catch royalties, so no further discussion is required. (An exception would arise for fisheries in which individual annual catches are highly unpredictable. Here catch royalties would probably be superior to royalties based on quota holdings. Weitzman (2002) discusses the relative merits of taxes and individual quotas under uncertainty.)

Next, for the case of royalties based on annual net revenue, let τ now denote the corresponding royalty rate. Then

$$PV_{i,\tau}(Q_i) = \frac{1-\tau}{r_i}[pQ_i - c_i(Q_i)]$$

and the corresponding quota demand equation becomes

$$c_i'(Q_{i,\tau}) = p - \frac{r_i m_\tau}{1-\tau}$$

For example, if all fishermen have the same costs and discount rates, we obtain, by the previous calculation,

$$m_\tau = (1-\tau)m$$

Net revenues are reduced by proportion τ, and so are quota values. It is easy to show that we have in general

$$Q_{i,\tau} = Q_{i,0}$$

i.e., royalties based on net revenues have no effect on the post-market distribution of catch quotas.

IFQs, ITQs and Property Rights

One often encounters the statement that IFQs are a form of property rights in the fishery resource. Just as the owner of a piece of real property, such as a house, a farm, or a mine, will have every incentive to use that property in an economically rational manner, so (it is asserted) will IFQ owners be motivated to efficiently harvest the resource, *and to protect and conserve the resource stock*. Therefore, such so-called rights-based fishing systems are said to be both necessary and sufficient for effective resource exploitation and conservation in fisheries and elsewhere (see, e.g., Neher et al. 1989).

But do IFQs or ITQs actually encourage resource conservation necessary for enhancing the value of the quotas? How could we model this question? Let us imagine an IFQ-based fishery that operates on two fishing grounds, A and B. Unknown to the resource managers (but perhaps known to the fishermen), trawling activities in ground A gradually destroy the bottom topography, reducing local productivity. This is not a problem on ground B. At present, fishing is more profitable on ground A than on B. The long-term optimal strategy, we suppose, is to fish ground B more intensively than A, thereby preserving overall resource productivity. Some of the fish from ground A will migrate to B if fishing reduces the stock there.

Clearly under TAC-regulated open access (without IFQs), fishermen will tend to concentrate initially on ground A. Will IFQs automatically reverse this trend? The fishermen are fully aware that using A will lead to an ultimate reduction in the TAC, and hence in their individual quotas. Would IFQs or ITQs automatically reverse the situation, so that the fishermen would then be self-motivated to use the optimal strategy?

Without constructing an explicit model (which could be done), we recognize this as a typical game-theory situation. If all fishermen avoid A they all gain accordingly (and the market value of their quotas will remain high). But a single defector who fishes A will increase his annual revenue (by catching his quota more easily). The same argument applies to all fishermen, so that all will choose to fish on ground A. The IFQ system does nothing to interfere with this standard game-theory situation.

Two counter-arguments are immediately suggested. The first is that this is merely a management failure: the managers should close area A, or otherwise regulate fishing there. Right, but that is not the point. We are asking whether IFQs are sufficiently like property rights to engender optimal resource conservation. The answer is negative.

Second, one might argue that, because fishing area A will reduce the value of their quotas, fishermen will be motivated *to agree* to avoid area A. Right again, but it remains true that the IFQ system does not imply (as full property rights would do) that the fisherman's individually optimal fishing strategy will favor resource conservation, i.e., avoiding area A. Nevertheless, since the quota owners have exclusive future access to the resource, as a group (or so we are assuming), they benefit from cooperative action. They may therefore agree to avoid area A. To be successful, such an agreement would have to include some form of coercion to prevent defection. Perhaps the quota owners would ask the management agency to regulate the use of area A.

This example suggests that although IFQs may not directly motivate fishing strategies that would benefit the group of fishermen as a whole, the existence of an IFQ system may encourage cooperative behavior. The difference between an IFQ system and a non-IFQ system is that with IFQs, each fisherman owns a valuable long-term asset which he is anxious to protect and enhance, through group cooperation as well as his own optimal strategy.

In this respect, IFQs do seem to have some of the features of private property rights, though perhaps not to as great a degree. After all, home owners' rights are also limited or restricted in various ways, which are designed to ensure cooperative behavior intended to prevent undesirable externalities, or encourage beneficial ones.

Individual Effort Quotas

In a fishery operating under individual catch quotas, reliable up-to-date information on each quota holder's daily catches is absolutely essential to the success of the IFQ system. While the necessary off-loading and monitoring of catches at a few specific locations may be feasible for local, coastal fisheries, this may prove difficult for foreign or distant-water fleets.

In such cases individual effort quotas (IEQs) may be superior to catch quotas (IFQs). The current location of each fishing vessel can be accurately determined via satellite, so that the seasonal effort (vessel days) of each vessel, on a given fishing ground, is readily determined. Provided that the relative capacity of each vessel is fairly accurately known, IEQs would have similar bioeconomic implications to IFQs. However an incentive for capital stuffing, especially if it can be concealed from the managers, will always exist under an IEQ system. (Recall that there is no such incentive in a properly monitored IFQ system.) Consequently it seems likely that actual fishing capacity would gradually increase under a system of effort quotas. If so, the fishery might converge on the classical bionomic equilibrium, combining overfishing and overcapacity.

An effort-quota system would provide managers with no direct information on annual catches. Presumably the total annual effort quota E_{total} would be determined from the estimated TAC together with an estimated value for catchability. Whether the actual annual catch is anywhere near the TAC might never be known. Of course the same might be said for catch-quota systems, in the event that actual catches differ greatly from reported catches. This merely emphasizes the need to monitor accurately.

In short, effort quotas are more indirect in their effects than catch quotas. Effort quotas may be easier to administer than catch quotas, but they may also be more subject to management error.

Variability in Catch Rates

Another important difference between catch quotas and effort quotas stems from variability in catch rates. The deterministic equation

$$h = qEx \qquad (4.81)$$

that we have assumed until now, should realistically include a random term:

$$h = qEx \exp(\epsilon - \sigma^2/2) \tag{4.82}$$

where ϵ is a normal random variable, with mean 0 and variance σ^2. Here h and E refer, for example, to daily catch and effort of a given vessel.

Suppose, for example, that the vessel has an annual catch quota Q. It continues to fish until the quota is taken or the season ends, whichever comes first. The annual effort is then random: lucky fishermen (or skilled ones) quickly fill their quotas, unlucky ones keep on fishing. In this situation, annual revenues are predictable, but annual costs are random.

Similarly, for a vessel that has an annual effort quota, annual costs are predictable while annual revenues are random. It is not clear without developing a specific model whether annual net revenue would be more variable under a catch quota or an effort quota. Intuitively it seems likely that vessel owners might prefer predictable revenues and variable costs, as in an IFQ program, to the reverse situation typical of an IEQ program. However, fishermen do have a reputation as risk takers. Some may even dislike any form of individual quota system because it reduces the chance of a big season. Those who really prefer a career with low or negative incomes in most years, with an occasional bonanza, to a more regular positive annual income, will probably never be happy with IQs. You can't satisfy everyone.

Stock Fluctuations and Quota Shares

The tacit assumption so far has been that the total annual catch quota, or TAC, remains constant from year to year. This is unrealistic for most, if not all actual fisheries. Annual recruitment is often highly variable, requiring an annual adjustment to the TAC. Also improvements in stock estimation may indicate that the present TAC is too high, or too low, again requiring TAC adjustments. Thus individual quotas also need to be adjustable from one year to the next.

To achieve adjustability, fishermen are often awarded quota shares rather than fixed quotas. (The acronym IFQS—Individual Fishing Quota Shares—is sometimes used for this case.) The annual TAC then determines the quantity of fish associated with one quota share for the current season, although this could be further adjusted during the season in some cases.

It is sometimes said that fluctuations in quotas, particularly reductions from previous levels, are undesirable. However, most fish populations are inherently variable, so that some degree of flexibility in catch quotas seems inevitable. While quota reductions in a fishery at bionomic equilibrium (regulated or otherwise) may be especially undesirable, since fishermen may then wind up with small or negative annual returns, the problem should be much less severe under an IFQS system. The fishermen should realize that the long-term viability of their fishery may require occasional sharp reductions in catches. The dissipation of profits having been largely eliminated by the IFQS system, few fishermen should experience severe financial difficulties in years where the quota is below average. It is conceivable that insurance could be purchased to cover such exigencies. The insurance would pay off whenever the TAC fell below some specified level.

Setting the Annual TAC

These considerations raise the question of how the annual TAC should be set in the first place. In order to limit the need for frequent adjustments to the TAC, it would seem desirable to set TACs well below any estimated average MSY level. As explained fully in Chapter 3, it is usually not economically desirable to set the TAC at zero whenever the current biomass is below the long-term target. Rather, setting the TAC at or near the long-term equilibrium level will often be appropriate unless this seems likely to further deplete the stock.

To what extent should economic considerations affect the calculation of the long-term average TAC? The simple deterministic bioeconomic models of Chapters 2 and 3 would seem to indicate a strong economic component to TAC (and target biomass) calculations. However, because of the dominant effect of biological uncertainty in most fisheries, actual determination of TACs should probably be based primarily on biological considerations—sustainability in particular. As just noted, economic considerations do suggest that recovery phases for depleted stocks should usually involve less extreme quota reductions than might otherwise seem appropriate. We discuss these questions further in Chapter 5.

4.4 Quota Auctions

Another method for obtaining government revenues from an IFQ (or IEQ) fishery is by auctioning quotas.

Assume that individual quotas are auctioned to the highest bidders, with a total annual quota Q. The quotas have a term of T years. If the ith fisherman puts in a bid for Q_i quota units, at price w_i, his net present value (if he obtains the quota) will equal

$$PV_{i,T}(Q_i) - w_i Q_i \qquad (4.83)$$

where $PV_{i,T}(Q_i)$ is given by Eq. (4.67).

We now make the idealistic assumption that all fishermen correctly predict the prices that other fishermen will offer. Then in fact all fishermen will make the same offer, $w_i = w$. Knowing the value of w, the ith fisherman will determine Q_i to maximize his net present value, Eq. (4.83). Hence we have

$$PV'_{i,T}(Q_i) = w \qquad (4.84)$$

$$\sum_{i=1}^{N} Q_i = Q \qquad (4.85)$$

exactly as in the case of tradeable quotas—see Eqs. (4.69) and (4.70). Thus under our present assumptions, the auction price w is the same as the market price m_T for quota trades. The auctioning of quotas thus has similar economic implications to the establishment of a quota market in an ITQ system. Total profits derived from the fishery are maximized under either system, at least if the fishermen all use the same discount rate r. There is, however, one fundamental difference between the two management systems. With quota auctions, most of the profits are garnered by the government, whereas in an ITQ system with free initial quota allocations the profits go to the original recipients.

Which system is superior? And is there some intermediate system, with profits being split between the government and the fishermen? Deciding who should reap the profits from a publicly owned fishery resource is a political question. I have stated my position earlier—at least a substantial portion of the profits should go to the ultimate owner of the resource, which means the public (Macinko and Bromley 2002). However, the fishermen also probably deserve to obtain a fair share of the profits. Also, unless fishermen receive some of the benefits obtained from managing the fishery, they will have little incentive to protect the resource or to retain profitability. Fisheries management has long been plagued by resistance on the part of the fishermen. Neither freely granted ITQs (or IFQs or IEQs), or auctioned quotas, are capable of splitting the resource profits. However, quota fees can do so.

Another way of sharing profits is provided by partial quota auctions

(Huppert 2006). Here part of the TAC is assigned to fishermen as their basic ITQs while the remainder is auctioned annually. If the fees attached to the basic ITQs are low, this method allows for profit sharing between the fishermen and the government. Also, annual fluctuations in the TAC could largely be encompassed by variations in the auctioned portion, in which case annual government income would be more variable than fishermen's incomes. These and other aspects of partial quota auctions are discussed by Huppert (2006).

4.5 Some Aspects of Multispecies Fisheries

Numerous papers have been published in recent years criticizing the practice of managing fisheries on the basis of single-species models (e.g. Pikitch et al. 2004). Obviously no fish population lives in isolation from other marine organisms—predators, prey, competitors, pathogens—but it is not entirely clear how to incorporate multispecies and ecosystem effects into management models. Adding economic components to the mix makes the problem even more complex. Nevertheless, since multispecies aspects of fisheries often have major economic implications, these need to be taken into consideration in setting management policies.

Some aspects of multispecies phenomena with important economic implications are:

1. Simultaneous harvesting of ecologically linked species, whether as separately targeted catches taken by different vessels, or as mixed catches taken by the same vessels. Examples include predator-prey systems (e.g., seals, cod and capelin, usually caught separately), and groups of coexisting groundfish species caught in the same trawls.
2. The problem of by-catches, in which substantial numbers of non-target species are inadvertently captured together with the target species. The by-catch may be retained because of its value, or discarded because of zero or low value. In some cases, regulations may require the discarding of valuable by-catch species for conservation reasons.
3. The gradual depletion or elimination of low-productivity (often large-size) species in a mixed-species fishery. Unless carefully regulated, such fisheries will progressively catch mainly smaller species, a process that has been termed "trophic overfishing."
4. Problems of habitat degradation resulting from fishing activities, such

as bottom trawling (Kaiser et al. 2002). An extreme example is the destruction of coral reefs from dynamiting or poisoning to force reef fish into the open. Activities of this kind permanently destroy the resource, and would only occur in a lawless, unregulated fishery, or perhaps under high rates of discount.

Theoretically speaking, there is no difficulty to write down the equations of a general multispecies, dynamic bioeconomic fishery model, extending models of Chapter 2. Also, computer simulation models can mimic the complexities of ecosystems, but few such models have included detailed economic components. (See Walters and Martell 2004 for biological aspects of ecosystem models.)

A Predator–Prey Model

As an aid to discussion, let us consider the following predator–prey bioeconomic model:

$$\frac{dx}{dt} = G_1(x) - a_1 xy - h_1 \tag{4.86}$$

$$\frac{dy}{dt} = G_2(y) + a_2 xy - h_2 \tag{4.87}$$

where x and y are the biomass levels of the prey and predator populations, respectively. The terms $\pm a_i xy$ model the predator–prey interaction. We assume independent fisheries for x and y, and suppose that

$$h_1 = q_1 E_1 x \text{ and } h_2 = q_2 E_2 y \tag{4.88}$$

Net revenue flows in the two fisheries are:

$$R_1 = (p_1 q_1 x - c_1)E_1 \text{ and } R_2 = (p_2 q_2 y - c_2)E_2 \tag{4.89}$$

We can now carry out similar calculations to those of Chapter 2, identifying and comparing open-access bionomic equilibria with optimal target equilibria. Optimal dynamic harvest strategies could also be determined, although this is quite complicated (some details are in Clark, 1990, ch. 10). For example, the present-value maximization problem is now

$$\text{maximize} \int_0^\infty e^{-\delta t}\{(p_1 - c_1(x))h_1(t) + (p_2 - c_2(y))h_2(t)\}dt \tag{4.90}$$

subject to Eqs. (4.86) and (4.87), and with $c_1(x) = c_1/q_1 x$, etc.

This optimization problem has an equilibrium solution (x^*, y^*), determined by the equations

$$\frac{\partial R}{\partial x} = \delta(p_1 - c_1(x))$$
$$\frac{\partial R}{\partial y} = \delta(p_2 - c_2(y)) \tag{4.91}$$

where $R(x, y)$ denotes sustainable economic rent:

$$R(x, y) = (p_1 - c_1(x))(G_1(x) - a_1 xy) + (p_2 - c_2(y))(G_2(y) + a_2 xy) \tag{4.92}$$

These equations are the two-species analog of Eq. (2.63) for the single-species model.

What insights, if any, can be obtained from this model? First, as far as open-access overfishing is concerned, we have

$$x_{OA} < x^*$$

but

$$y_{OA} \text{ may be } < y^* \text{ or } > y^*$$

(Here $x_{OA} = c_1/p_1 q_1$ and $y_{OA} = c_2/p_2 q_2$ are the bionomic equilibrium stock levels for the two fisheries.) The open-access fishery on the prey species x does not take account of the positive effect that prey have on the predator population, so it will tend to overexploit the prey species for this reason, as well as for the usual reasons discussed in Chapter 2. Fisheries managers are well aware of the need to limit the catches of commercially important prey species in order to protect the food supply of valuable predator species.

Likewise, the foraging requirements of sea birds and marine mammals need to be recognized in setting quotas for their prey populations. We may be a long way from estimating a shadow price for Northern Gannets in Atlantic Canada, but this does not imply that management of capelin stocks around Bonaventure Island should overlook the requirements of the famous Gannet colony located there.

Whether the predator species y will be overfished or underfished in an open-access fishery depends on the relative prices of the two species. Removing predators is beneficial for the prey population (according to our two-species model), but if the predators have little or no commercial value they will be lightly harvested or unharvested. Fishermen often complain, for example, that marine mammals eat too many fish, and should be culled. Mammal watchers may likewise complain that the fishermen catch too many fish, and should be regulated. Standard techniques for the valuation of non-market benefits could be used to estimate preservation values for species impacted by commercial fisheries.

This example indicates that, contrary to the impression that might result from studying simple models, ITQ systems alone may not motivate fishermen to favor socially optimal management strategies. Externalities, for example in multispecies fisheries, may be important to society at large, but not to the fishing industry to the same degree. In general, such externality conflicts will be most severe in unregulated open-access fisheries (for example, deep-sea fisheries). In some cases, participants in IFQ or ITQ fisheries have exhibited willingness to cooperate in protecting non-target species and ecosystems (see Chapter 6). Any tendency for individual captains to infringe upon agreed strategies can be, and often is, countered by the use of on-board observers paid by vessel owners themselves.

A Mixed-Species Model

A mixed-species model may be formulated as follows:

$$\frac{dx_i}{dt} = G_i(x_i) - q_i x_i E \quad (i = 1, 2, \ldots, n) \tag{4.93}$$

This assumes that the n species are ecologically independent (otherwise $G_i(x_i)$ would be replaced by $G_i(x_1, \ldots, x_n)$), and that a single effort level E applies to all species in the fishery. Fleet revenue flow is

$$R = \left(\sum_{i=1}^{n} p_i q_i x_i - c \right) E \tag{4.94}$$

Bionomic equilibrium is characterized by $R = 0$ and $dx_i/dt = 0$ ($i = 1, 2, \ldots, n$)—a system of $n + 1$ equations in $n + 1$ unknowns. We can solve this system explicitly if $G_i(x_i) = r_i x_i(1 - x_i/K_i)$. From $dx_i/dt = 0$ we obtain $G(x_i)/q_i x_i = E$, which reduces to

$$x_i = \begin{cases} K_i(1 - \frac{q_i E}{r_i}) & \text{if } q_i E < r_i \\ 0 & \text{otherwise} \end{cases} \tag{4.95}$$

From $R = 0$ we obtain

$$\sum_{x_i \neq 0} p_i q_i x_i = c$$

(or else $E = 0$). Combining this with Eq. (4.95), we conclude that, at equilibrium

$$E = \frac{\sum p_i q_i K_i - c}{\sum p_i q_i^2 K_i / r_i} \qquad (4.96)$$

where the sums are taken over $x_i \neq 0$.

Equation (4.95) implies that in the unregulated mixed species fishery:

species i will be eliminated if $\quad \dfrac{r_i}{q_i} \leq E \qquad (4.97)$

Recalling that $q_i E$ is the fishing mortality on species i, we see that this result is already known from Chapter 2. The difference here is that, because the fishery can continue viably even when some species have been fished out, there is no automatic "brake" on overfishing such species. This contrasts with the single species model (with Schaefer catch equation) of Chapter 2, in which $x_{OA} = c/pq$ is positive.

Here is a numerical illustration, for $n = 2$ species.

$$K_1 = K_2 = 10,000 \text{ tonnes}$$
$$r_1 = 0.2, \quad r_2 = 0.5 \text{ /yr}$$
$$q_1 = q_2 = .01/\text{SVU yr}$$
$$p_1 = 0, \quad p_2 = \$1,000/\text{tonne}$$
$$c = \$50,000/\text{SVU yr}$$

We have

$$\frac{r_1}{q_1} = 20 \text{ SVU}, \quad \frac{r_2}{q_2} = 50 \text{ SVU}$$

and, from Eq. (4.96)

$$E = 25 \text{ SVU}$$

Thus species 1 will be exterminated in this two-species fishery. In this example, species 1 is an economically worthless by-catch, the loss of which makes no difference to economic yield.

Note, however, that the same outcome occurs even if species 1 is valuable. For example, retaining the same parameter values, let us change p_1 to \$10,000/tonne. Species 1 is still eliminated in the open-access mixed fishery, even though this greatly reduces the profitability of the fishery.

We can also calculate the effort level E_{MEY} (maximum economic yield) that maximizes the sustainable economic rent:

$$\underset{E}{\text{maximize}} \left(\sum_i p_i q_i x_i - c \right) E \qquad (4.98)$$

$$\text{subject to} \quad q_i x_i E = G(x_i) \tag{4.99}$$

By substitution from Eq. (4.95) and differentiation, we obtain

$$E_{\text{MEY}} = \frac{\sum p_i q_i K_i - c}{2 \sum p_i q_i^2 K_i / r_i} \tag{4.100}$$

(Again the sums are taken for $x_i \neq 0$.)

For the second case of our numerical example we obtain $E_{\text{MEY}} = 10$ SVU. Therefore neither species would be eliminated under a management system based on MEY. (The dynamic optimization problem, i.e. with positive discounting, is probably beyond analytic solution techniques.)

Is it possible that an economically optimal (MEY) fishing strategy would actually eliminate some valuable species in a mixed-species fishery, as modeled here? Indeed this is possible. Suppose $p_1 = \$1,000$/tonne and $p_2 = \$10,000$/tonne in our example. Equation (4.100) gives $E_{\text{MEY}} = 23.75$ SVU, so that species 1 would be eliminated under optimal fishing.

An important example of this kind occurs in the British Columbia salmon fisheries. Salmon runs often contain a mixture of several genetically distinct stocks, either different species or different substocks of a single species. The fishery inevitably catches a mixture of the stocks. The result has been that over time many minor substocks have been almost eliminated. Cutting back the salmon TAC sufficiently to protect these substocks would have a devastating economic effect on the fishery. While fine tuning of the timing of the fishery can apparently reduce the problem to some degree, it seems unlikely that complete protection of all 3,000 stocks of salmon in B.C. will ever be achieved.

Returning to the mixed-species model, what management options exist to prevent the elimination of low-productivity species? As far as the model is concerned, the only option would be a reduction in effort. In practice, several additional options may exist, including the use of closed seasons or areas to protect vulnerable stocks. Also, changes in the design of fishing gear may reduce the mortality of some species.

Another management option is the use of regulations that prohibit the retention of vulnerable species in the catch. Although such species may be still caught and then discarded, the regulations may affect fishermen's behavior, causing them to avoid areas with high local abundance of the prohibited species.

Yet another option is the use of individual vessel by-catch quotas. If combined with ITQs for target species, this system can successfully alter fishing patterns while retaining profitable operations (Grafton et al. 2004).

A Source–Sink Model

Most single-species models treat the fish population as a single homogeneous unit (possibly with age structure—see Sec. 2.7). In actuality a given population may be spread out over a large area, often covering thousands or even millions of square kilometers. This spatial distribution has important implications for population dynamics, and also for fishery management (Sanchirico and Wilen 1999, 2005; Crowder et al. 2000; Holland 2004; Wilen 2004). Here I consider a simple source–sink model, in which the source sub-population inhabits the breeding grounds, and exports adult fish to the sink population.

Let $x_1(t)$ and $x_2(t)$ denote the source and sink populations, respectively. The model equations for the unfished system are

$$\frac{dx_1}{dt} = rx_1(1 - x_1/K) - a(x_1 - x_2) \qquad (4.101)$$

$$\frac{dx_2}{dt} = a(x_1 - x_2) - mx_2 \qquad (4.102)$$

Here a is a constant, specifying the net migration rate from source to sink.

The natural equilibrium (x_1, x_2) is obtained by setting $dx_1/dt = dx_2/dt = 0$ and solving for x_1, x_2. The result is

$$x_1 = \frac{aK}{r}\left(\frac{r}{a} - \frac{m}{m+a}\right) \qquad (4.103)$$

$$x_2 = \frac{a}{m+a}x_1 \qquad (4.104)$$

To have $x_1 > 0$ (a viable population) we require that $r/a > m/(m+a)$, and this can be rewritten as

$$\frac{1}{r} < \frac{1}{a} + \frac{1}{m} \qquad (4.105)$$

This inequality says that the intrinsic growth rate (r) must be large enough to make up for emigration (a) and natural mortality (m) of the sink population. If this condition holds, then (x_1, x_2) is a positive stable equilibrium for the source–sink model.

Next we consider the effect of constant-rate harvesting of either or both sub-populations:

$$\frac{dx_1}{dt} = rx_1(1 - x_1/K) - a(x_1 - x_2) - F_1x_1 \qquad (4.106)$$

$$\frac{dx_2}{dt} = a(x_1 - x_2) - mx_2 - F_2x_2 \qquad (4.107)$$

These equations can be written in the form

$$\frac{dx_1}{dt} = (r - F_1)x_1\left(1 - \frac{r}{r - F_1}\frac{x_1}{K}\right) - a(x_1 - x_2) \quad (4.108)$$

$$\frac{dx_2}{dt} = a(x_1 - x_2) - (m + F_2)x_2 \qquad (4.109)$$

The viability condition, Eq. (4.105), now becomes

$$\frac{1}{r - F_1} < \frac{1}{a} + \frac{1}{m + F_2} \qquad (4.110)$$

If this condition is not met then fishing will drive the combined population towards extinction.

For example, if $F_1 \geq r$ the source population will tend towards extinction. This is the same as for the logistic model discussed in Chapter 2. In the present model, however, extinction can occur also for $F < r_1$, as a result of harvesting the sink population too intensively.

Perhaps the most interesting example arises if $1/r < 1/a$ (i.e., $r > a$), in which case no amount of harvesting of the sink population alone can wipe out the entire resource. But if the source population is also harvested, a population crash becomes possible. This scenario fits the history of the cod fishery in the Western Atlantic. Inshore cod stocks supported a highly productive fishery for many centuries, but the stocks crashed unexpectedly in 1991 after a couple of decades of fishing the offshore breeding stocks.

We can also add a bit of economic realism to our source–sink model. We consider an open-access fishery, and assume Type I CPUE profiles (Sec. 2.27) for both sub-populations. Specifically, suppose that $h_i = q_iE_i$ ($i = 1, 2$) where E_i denotes effort for sub-population i. Net revenue flow for the i-fishery is given by

$$R_i = ph_i - c_iE_i = (pq_i - c_i)E_i \qquad (4.111)$$

Thus for each sub-population there is a threshold cost price ratio $\theta_i = q_i$ such that fishing is economically viable if and only if

$$\frac{c_i}{p} < \theta_i \qquad (4.112)$$

Suppose, for example, that the sink population x_2 is more accessible than the source population. The inequality (4.112) may hold for $i = 2$ but not for $i = 1$. Under open-access conditions the sink population will be fished down (almost) to zero. Nevertheless the fishery is sustainable (assuming that $r > a$), because the source population remains unfished. According to the model equations (4.16), (4.107) we have (with $x_2 = 0$)

$$h_2 = ax_1 = a(r - a)K/r$$

The situation is stable—provided that the source population remains unfished. But what if economic conditions change? For example, either an increase in price p, or a decrease in effort costs c_1 can flip the threshold condition of inequality (4.112). If so, the source population suddenly becomes profitable, and this spells doom for the entire open-access fishery, which will rapidly collapse. The history of such a fishery will show a long period of sustained fishing, followed by a sudden transition to a state of collapse.

Examples that seem to fit this scenario include, once again, Canada's Atlantic cod and other ground-fisheries, and the spring-spawning herring fishery in the North Sea (Bjørndal et al. 2004). The latter stock, which had supported sustained fishing for more than 100 years, collapsed from over 10 million tonnes to a few thousand tonnes in the late 1960s, as the exploited area expanded to include most of the stock's distribution. A fishing moratorium established in 1970 was followed by recovery of the herring stocks, and fishing was resumed by 1990. See Mullon et al. 2005 for discussing of the dynamics of fishery collapses throughout the world.

Our source–sink model suggests that a marine reserve designed to protect the source sub-population might be a good idea, in terms of hedging against uncertainty in TAC specifications. Bioeconomic spatial models of fish population dynamics have been developed and analyzed by Sanchirico and Wilen (2001; 2005).

What is the optimal harvest strategy for a source–sink population? We begin by determining the MSY solution. Setting $dx_1/dt = 0$ and $dx_2/dt = 0$ in Eqs. (4.108) and (4.109), and then adding the two right-side expressions, we obtain

$$Y = F_1 x_1 + F_2 x_2 = rx_1(1 - x_1/K) - mx_2 \qquad (4.113)$$

To maximize this expression one should reduce x_2 to zero, and keep $x_1 = K/2$. That is, the source population should be held at its own MSY level, and the sink population should be heavily exploited to overcome losses from natural mortality. But is this theoretical result of any practical significance?

At this stage the limitations of our simplistic source–sink model become apparent. In most fish populations, adult fish from the sink area migrate back to the source area to spawn. To include this feature, we would need to include age structure and age-dependent migration and reproduction in the model. (In Atlantic cod the inshore stocks consist mainly of juvenile fish.)

So what's the point of discussing the simple source–sink model here? The point is just to indicate that there may be many important complexities that are not often included in management models. Modeling such complexities, for example by using simulation methods (Walters and Martell 2004) may reveal unexpected system properties, but how to adapt management strategies to take these aspects into account may be far from obvious. Which of countlessly many conceivable biological complexities are relevant to management will probably remain virtually unknowable, except perhaps in retrospect (McAllister and Kirchner 2002).

Some examples of factors that have been suggested as affecting the collapse and non-recovery of marine fish populations are:

1. Habitat destruction (Kaiser et al. 2002)
2. Depensation (Liermann and Hilborn 2001)
3. Fish behavior (Myers et al. 1997)
4. Ecosystem shifts (Walters and Martell 2004)
5. Spatial structure (Berkeley et al. 2004)
6. By-catches.

Would the introduction of an ITQ-based system of management help to resolve such system-related difficulties? Assuming that quota owners are motivated to protect and enhance their assets, there should at least be a strong incentive to support necessary research, and also to devise and employ robust harvesting strategies that can reduce the risk of catastrophic collapses. Some of the case studies discussed in Chapter 6 support this claim.

4.6 Compliance

Compliance is one of the most important issues in fisheries management; illegal fishing and failure to report catches are common occurrences both within and beyond EEZs (Kuperan and Sutinen 1998; Charles et al. 1999; Sumaila et al. 2006). As has been stressed repeatedly in this book, fishermen are almost always motivated to increase their effort, so as to attain greater catches. Individual catch or effort quota programs directly counter this motivation by, in fact, outlawing larger catches or effort levels. But outlawing an activity is not the same as making it undesirable—to the contrary, it may render the activity more attractive.

In the case of fisheries, full compliance will benefit all participants, who will enjoy the sustained profits thereby produced. The question is, how to ensure compliance in spite of the fact that individual fishermen will always be motivated not to comply. Here I will limit the discussion to individual catch quotas, although other compliance issues may also be important in practice.

An IFQ system will fail completely unless the individual quotas are strictly enforced. Thus some virtually ironclad system of monitoring actual catches as they are taken, is a necessity. Typically this is accomplished by requiring that all catches be landed at one or more specific locations, where government inspectors monitor and record each vessel's landings. Offloading catches at other locations is treated as a major offense.

Another approach that is sometimes used involves on-board observers, who monitor and record catches (and other aspects of fishing activities) at sea. Observers may be useful for scientific data collection and analysis, but cannot usually be expected to deal with compliance issues. An on-board government observer who checks and reports on noncompliance activities is usually regarded as a spy, and may be treated accordingly. However, if observers are hired and paid by the vessel captains themselves, these difficulties may be averted.

One way around the problem of monitoring and enforcement of catch quotas is the use of effort quotas. For example, the annual effort of a specific vessel may be defined as total hours spent on the fishing rounds, multiplied by a coefficient related to the vessel's fishing capacity. Monitoring of time spent on the fishing ground may be achieved fairly easily by using on-board satellite transponders. (This method can also be used to ensure that the vessel is not fishing in closed or protected areas.) Time away from port is a simpler, if less accurate, measure of annual effort.

Enforcement of effort quotas may be much easier than for catch quotas, but the cheating incentive will encourage vessel owners to understate or conceal the true fishing power of their vessels. Thus actual total annual effort may exceed nominal effort, with the result that actual catches are greater than the intended level. The discrepancy may grow larger over time, as the fishing power of vessels gradually increases. Continuing inspection and re-assessment of capacity of licensed vessels is required to counteract these developments.

Resolving compliance issues is an essential component of any individual-quota-based management system. Cooperation of the fishermen themselves in designing a compliance system, and helping to ensure that it is enforced may also be essential. Each fisherman must be satisfied that everyone is following the same rules. This may sound simple enough, but the facts of life are that in many fisheries, breaking the rules has long been the norm.

Certain fishing regulations may not benefit the fishermen themselves. By-catch restrictions are one example. Fishermen holding quotas for a given species may inadvertently catch other species; indeed they may target other species if they are more valuable. In addition, some techniques (e.g. long lines) may incidentally kill large numbers of other organisms, including sea birds and marine mammals. Regulation designed to control these incidental kills may increase the costs of fishing. If so, fishermen will be motivated to ignore the regulations. However, this motivation will usually be much weaker under an IFQ system than under TAC-type regulation alone.

4.7 Whose Fish?

The theme of this book is that marine fisheries can be managed for sustained profitability, although doing so may be more demanding than is generally recognized (Caddy and Seijo 2005). If this is correct, the question arises as to who is entitled to share in the profits.

Until recently, few fisheries have succeeded in generating sustained profits. Temporary profits from the initial development of the fishery accrue to the first fishermen to enter the fishery; indeed, these initial profits often motivate overexpansion, which leads to overfishing, possibly to the point of short-term bionomic equilibrium. Sustained operating profits are then either zero, or in any event far less than the optimum level.

At this stage, management programs have typically been concerned with stock recovery and protection. Where successful in these terms, such management programs have often enticed further expansion of the fishing fleets, as fishermen begin to realize that profits are once again available, even if only temporarily. Thus a new, regulated bionomic equilibrium emerges, again with zero or suboptimal sustained profits. The regulated equilibrium, characterized by overcapacity, abbreviated fishing seasons, and severe management difficulties, is the situation that exists in many of today's fisheries.

The remedy suggested here, individual fishing quotas, is currently in use, or under consideration, in several countries. Provided that compliance issues can be successfully resolved, IFQs or similar management systems are capable of generating substantial, near-optimal profits from the fishery. Who should receive these profits?

In a situation where potential profits have historically always been allowed to dissipate, asking this question may seem senseless. But the question does need to be addressed during the design stage of any new management system.

The fishing industry may argue that 100% of profits ought to be retained by the industry, through freely granted fishing rights. If this policy is adopted, the government then faces the problem of allocating profits between various segments of the fishery. Indeed, allocation problems currently dominate many, if not most fisheries management programs. In extreme cases, some groups of fishermen may resort to the courts for relief from unfair allocation decisions.

The polar alternative is for the government to sell fishing quotas to the highest bidders, letting the market resolve all allocation issues. In this case most of the profits will be received by government (unless collusion among bidders takes place). Is this a feasible, or a fair solution?

The principal problem with an allocation system that retains all or most of the profits for the government, is that fishermen then have little incentive to protect the resource, since they have no financial stake in it. Not only will each fisherman wish to exceed his quota, but he will also not be much concerned if others do so, too. The "us versus them" attitude that currently plagues many fisheries will persist under a rent-removing system of management.

Although it was argued in Sec. 4.4 that IFQs and ITQs are incomplete property rights, such quotas will nevertheless be considered as financial assets by their owners. The owners will favor management measures that maintain or enhance the value of their quotas, such as protecting the resource stock, hedging against resource shocks and management errors, and so on. The greater the asset value, the stronger will be the support for such measures.

This does not necessarily imply, however, that the ultimate resource owner, the public (McRae and Munro 1989), must grant all resource income to the fishing industry. Perhaps a 50-50 split, after all management costs have been assessed, would be considered fair. Certainly the fashion among some economists to claim that distributional matters are irrelevant should not be taken seriously.

Similar arguments pertain to two other situations. First, if quotas are transferable, and if some fishermen sell their quotas to non-fishermen, who subsequently lease them back to active fishermen on an annual basis, then the lessees will have little incentive for resource conservation. Similarly, if a nation sells annual quotas to foreign fishermen, the latter will again have little incentive for conservation. These examples are all instances of the so-called principal–agent problem (Clarke and Munro 1987). The objectives of principal and agent differ—here, the principal favors conservation, which increases the value of his asset (the fishing quota), but the agent favors short-term revenues, since conservation does not increase his welfare. For the case of ITQs, the implications are that quota transfers to non-fishermen are undesirable, and also that a fair system of revenue sharing between the resource owner and the fishermen is needed. It is less clear how foreign fishermen could be induced to willingly protect the resource; perhaps this helps to explain why most nations have excluded foreigners from their 200-mile fishing zones.

4.8 Summary of Chapter 4

We began this chapter with a discussion of royalty charges as a method of fishery management. Royalties are often called taxes, but this word carries a connotation of unfairness—why should fishermen be "taxed" twice? Royalties are special user charges, paid only by those who profit from harvesting a publicly owned resource.

The basic Gordon–Schaefer model suggests that catch or effort royalties alone can prevent biological overfishing by altering the bionomic equilibrium stock level x_{OA}. This prediction, however, is model dependent, requiring a linear (Type II) or sublinear (Type III) CPUE profile. Since most fisheries probably have supralinear (Type I) CPUE profiles, the ability of royalties alone to control overfishing is limited. Output controls (TACs) are usually required in addition to royalties.

An alternative approach to catch royalties uses individual catch quotas (IQs, or IFQs). Many discussions treat catch royalties and individual catch quotas as an either-or choice. As we argue later in this chapter, however, effective fishery management requires both individual quotas and royalties.

Two management strategies that have often been used to control overfishing and overcapacity are total, non-individualized catch quotas (TACs) on the one hand, and limited licensing on the other. TACs prevent overfishing, thereby potentially preserving profitability of fishing. But this very profitability attracts entry to the fishery resulting in overcapacity and the derby fishery (modeled in Chapter 2). License limitation, an attempt to overcome this problem of excess capacity, has difficulties of its own, particularly capital stuffing.

An associated management initiative, a buy-back program designed to reduce excess capacity, was discussed fully in Chapter 3. Our conclusion there was that buy-backs tended to be costly and of limited effectiveness. The remainder of Chapter 4 discusses a different management approach, based on individual quotas.

Under an IFQ (individual fishing quota) system, each fisherman has a fixed annual catch quota for the species in question. The sum of the individual quotas equals the annual TAC. Related systems include IVQs (individual vessel quotas) and EAs (enterprise allocations). In practice an IFQ system is usually implemented as an IFQ-share (IFQS) system; this allows for necessary adjustments to the annual TAC.

Intuitively, IFQs change the rules of the competitive (non-cooperative) fishing game. An individual fisherman can no longer increase his share of the annual catch by employing high-intensity effort, or by expanding his vessel's capacity. Of course the motivation underlying such behavior still exists, but the assumption is that the individual quota system will be rigorously enforced. If so, the IFQ system effectively alters the fisherman's incentives.

Our model of the derby fishery (Chapter 2), modified to include IFQs, predicts that the IFQ system will in fact transform the fishery into the optimal profit maximizing exploitation mode. Rather than participating in a competitive race for fish during a brief open season, IFQ fishermen are motivated to maximize their income by employing economically efficient daily effort levels. Also, they are not motivated to undertake capital stuffing.

However, if the individual quotas are transferable (ITQs), some fishermen may wish to buy up quota units from other fishermen. A quota market will then develop in which the price of quotas is determined by balancing the supply of and demand for quotas. In order for transfers to occur, some fishermen must value quotas more highly than other fishermen. We showed that the value of a unit of quota to fisherman i depends on his marginal cost of effort $c_i'(E_i)$, and also on his personal discount, or time preference, rate r_i.

Specifically, fishermen with low marginal costs, or with low discount rates, will value quota units more highly than fishermen having higher marginal costs or discount rates. It follows that ownership of the quotas will tend to become concentrated in the hands of efficient low-discount-rate individuals. If transfers are unrestricted, the quotas may be bought up by non-fishermen, who will then lease the quotas back to active fishermen on a short-term basis.

Owners of ITQs can be expected to favor management strategies that enhance the value of their quotas. Examples include conservative TACs, protection of spawning grounds, avoiding the capture of small fish, and so on. The case studies discussed in Chapter 6 support this prediction.

This logic does not apply, however, in the case of fishermen who lease their quotas from the owners. In this situation the active fishermen will have no incentive to conserve or enhance the resource, because they themselves will not thereby derive any benefits. It is therefore probably inadvisable to permit quota ownership to be transfered to non-fishermen. In addition it may be argued that the profits from fishing should be reserved for active participants, not outside entrepreneurs. But to counter this argument, it must be noted that at least some of the future economic benefits do accrue to the original quota recipients (presumably fishermen), whether they use the quotas themselves or sell them off for a quick gain. A difference in personal discount rates implies that both buyer and seller consider themselves better off than before the transaction.

In the end, however, a fisherman who is forced to lease a quota from its owner will be no better off than in an open-access, profit-dissipating fishery. The ITQ fishery may be highly profitable, but the continuing profits are enjoyed by the quota owners, not necessarily by the active fishermen.

Empirical evidence from existing ITQ fisheries (see Chapter 6) supports the hypothesis that such fisheries can indeed by highly profitable, although there are some doubtful cases. This raises two important questions. First, what justification is there for deciding that these profits, derived from a publicly owned natural resource, should entirely accrue to a select few, the original quota owners? ITQ-based management systems will probably not receive public approval unless the distribution of the economic rewards is seen to be fair. The ability of ITQs to encourage conservation-oriented strategies requires that the resulting benefits are enjoyed in part by the quota owners, but this does not imply that 100% of those benefits must accrue to quota owners. The ultimate resource owner—the general public—deserves an equal share of the benefits.

Second, when an ITQ system is under consideration for a particular fishery, what is to prevent an anticipatory overexpansion of capacity, especially if it is believed that future profits will be entirely granted to participating vessel owners? Governments need to be up-front in insisting that fisheries are no different from other resource industries based on public resources. All users will be required to pay fair royalties for the privilege of owning restricted rights of access to the resource. Economists, many of whom have deliberately glossed over these questions, need to adjust their models to account for them.

Chapter 4 concludes with a brief discussion of some more biologically detailed models of fish population dynamics, including a predator–prey model, a model of a multi-species trawl fishery, and a source–sink model. These models pinpoint various limitations of the single-species models used elsewhere in this book. The bioeconomic principles discussed for single-species models can be extended to include such biological details, but no attempt is made here to be encyclopedic. Empirical evidence (Chapter 6) again suggests that ITQ-based management systems have the potential for dealing effectively with real-world biological complexity.

5

Risk Assessment and Risk Management

Uncertainty is of paramount importance in fisheries (Smith 1994). Virtually everything about a specific marine population is poorly known; some aspects may be almost completely unknown, even for important species with long histories of fishing. Past and present stock abundance, and biological characteristics such as growth, natural mortality and reproductive rates, though important to management decisions, are usually highly uncertain. Likewise, many details of ecosystem structure and the potential consequences of severely reducing target species, are largely unknown—some would say unknowable.

It is surprising, therefore, that until quite recently uncertainty and risk have been all but neglected in formulating scientific advice for management. The traditional paradigm, Maximum Sustained Yield (MSY), which dominated science-based management for many decades, was eventually modified marginally (for example, to the $F_{0.1}$ criterion commonly used in the 1970s–80s) to allow for unspecified uncertainty. But how to deal explicitly with uncertainty remained itself uncertain. Fishery managers often seemed to be interested only in the scientists' "best estimates" of the current Total Allowable Catch (TAC), and this was sometimes over ruled on political or economic grounds, namely because the proposed TAC was too small to satisfy the needs of the existing fleet.

By the early 1990s several major fisheries that had been under intensive management suffered severe stock collapses. These failures prompted many scientists to re-examine the whole question of risk and uncertainty in fisheries management. The following quotes indicate the scope of new ideas:

"A key element of fisheries stock assessment is the uncertainty associated with estimates of current stock size and potential productivity. The uncertainty in assessment leads to risk in decision making." (Hilborn et al. 1993.)

"As stock assessment methodology has moved toward statistical procedures, increasing attention has been focused on estimating the uncertainty in assessment advice.... Expressing assessment uncertainty to managers is another matter, however, and there is, as yet, no standard approach in the presentation of advice with respect to uncertainty and risk...." (Rosenberg and Restrepo 1994.)

"... a major reason for the increasing use of "risk" in the [fisheries] literature appears to have been a desire on the part of fisheries scientists to improve their advice to those who manage fisheries." (Francis and Shotten 1997.)

" The absence of this notion of "risk analysis" in decision-making is a major weakness of current fisheries management systems." (Lane and Stephenson 1998.)

"Although both types of uncertainty [i.e., parameter and structural uncertainty] should be accounted for, relatively little attention has been devoted to developing formal probabilistic methods to account for structural uncertainties." (McAllister and Kirchner 2002.)

These statements indicate, first, a tendency to concentrate on stock-size and parameter uncertainty, while down-playing other kinds of uncertainty, and second, a concern that uncertainty considerations may not influence management decisions in a desirable way. We discuss these and other questions in this chapter.

How in fact should uncertainty be taken into account in fisheries management? Let us first break this question down as follows.

1. Why consider uncertainty?
2. How can uncertainty be represented and quantified?
3. What are the implications of uncertainty in exploited populations?
4. How should management decisions incorporate uncertainty?

Consider the first question, why consider uncertainty? The alternative, as noted above, is for decision makers to request only "best estimates" from the scientists. For example, the scientists might announce that their best estimate for the current stock biomass is $x = 250,000$ t, and the best estimate of maximum sustainable fishing mortality is $F = 20\%$ per annum. The decision makers multiply these numbers (this is a caricature), and set the TAC accordingly, TAC = 50,000 tonnes. Next year

Table 5.1. *Probability distributions for stock abundance (artificial data).*

Stock	Probabilities
< 100,000 t	5%
100,000–200,000 t	21%
200,000–300,000 t	41%
300,000–400,000 t	28%
> 400,000	5%

the scientists announce that the previous year's estimate was probably in error; they estimate current abundance at $x = 180,000$ t. They also revise their estimate of F_{max} to 15% per annum. This implies a TAC of 27,000 t. On the grounds that this is extreme, the decision makers set a TAC of 35,000 tonnes. The fishery records a catch of 37,000 tonnes, but the scientists suspect that unreported catches may have amounted to an additional 5,000 tonnes at least. And so on.

This caricature may not be all that fanciful. But the process is clearly a risky one, especially given the fact that stock assessments are almost inevitably biased upwards in the case of a declining stock (Walters and Martell 2004, p. 59). What could be done to improve the situation? Should the scientists instead produce a probability distribution for the stock abundance? Table 5.1 is an illustration. Statistical methods exist, and are sometimes used, to produce such distributions.

If $F = 20\%$ (with no uncertainty—an unlikely situation), what TAC should be chosen in this case? Clearly there is no way of answering the question as it stands. What are the management objectives? What risks are involved, and what is the managers' attitude to risk? If a certain TAC, say 50,000 t is set for the year, is there some way of monitoring the fishery during the year to see if the TAC should later be reduced? If so, should an initial TAC of 40,000 t be set, with the possibility of increasing it later on?

Note that, even for this caricature, including uncertainty in a single quantity raises important issues that would probably not come up otherwise. One of the main advantages of explicitly recognizing uncertainty is that managers are forced to take a more comprehensive view of the system they are managing. They may also have to explicitly decide what their management objectives are. At least, the managers will come to realize that science has its limitations, but that risks can be identified, quantified, and perhaps managed. This will be a major advance, perhaps even a revolutionary advance, over past management practice (Hilborn et al. 1993).

How, in fact, should the TAC be specified in the above example? First, the best estimate of x is 210,000 t, and this would imply a TAC of 42,000 t if $F = .2$. What's wrong with this? Is it a safe decision? Clearly this cannot be answered without additional information. Also, some general strategy involving risk management is required. For example, suppose that a strategy of maintaining all fish stocks at $\geq 60\%$ of their estimated unfished biomass, has been adopted. For the case of Table 5.1 the estimated unfished stock level is 1.2 million t. The TAC should then be set at zero—unless the general strategy allows for small positive TACs in such cases.

A more interesting example would occur if $x_{\text{unfished}} = 500,000$ t, say. Deciding on the TAC might then involve estimating the probability of certain undesirable outcomes, for different TAC strategies. Important uncertainties could be identified, quantified, and included in the analysis.

Decision analysis, a well-developed methodology for decision making under uncertainty, is being increasingly used by government agencies and private firms in many fields (Morgan and Henrion 1990). Although decision analysis has to my knowledge never been fully applied in a fisheries setting, there is no reason why this should not happen. Section 5.2 will briefly outline the methods of decision analysis.

Some aspects of decision analysis, such as Bayesian information theory, and stochastic simulations, are beginning to be used by fisheries scientists (e.g., McAllister and Kirchner 2002; Parma 2001). However, a full implementation of decision analysis at the management level still awaits the future; see Chapter 7.

It must be recognized that explicitly considering risk and uncertainty could greatly complicate the task of decision making. As Hilborn et al. (1993) say, decision makers can always come up with another "what if" question, the answer to which may require months of research. We briefly address these issues in Sec. 5.2, and make some suggestions for limiting the complexity of risk-management programs in fisheries.

5.1 Kinds of Uncertainty

The words *risk* and *uncertainty* have many different interpretations. Both words refer to a lack of complete information, and this can arise in two distinct ways. First, a given system can undergo random future changes, so that even if the current state of the system is known, future states cannot be predicted, although the probabilities of future states may be known. Second, certain charactistics of the system may simply not be known or understood at all. Even the past history of the system may be poorly known.

The term "risk" is sometimes used (as in "insurable risk") to refer to random future events having a known probability distribution, derived from previous experience. In this case future uncertainty is captured by a probability distribution, which is entirely empirical. The frequency interpretation of probability is fully appropriate to this situation.

In this terminology, all other forms of incomplete knowledge are classified as uncertainties. Uncertainties may also be quantified in terms of probability distributions, but these no longer have a frequentist interpretation. Rather, the probabilities are best thought of as degrees of belief. Bayesian methods (Sec. 5.2) are often used to estimate such probabilities.

This use of the terms risk and uncertainty, though common in Economics, differs from normal use in Science. In this chapter I will follow Francis and Shotten (1997), who make the following definitions:

Uncertainty is the incompleteness of knowledge about the states or processes (past, present, and future) of nature.

Risk is the probability of something undesirable happening.

In this terminology, the term "risk assessment" refers to the process of estimating the probabilities of certain undesirable events, and also estimating the costs of these events.

Given the above definitions, it is clear that an almost unlimited number of uncertainties exist in fisheries. It will be convenient to first classify fishery uncertainties as Biological, Economic, or Management uncertainties.

Biological uncertainties consist of (cf. Francis and Shotten 1997):

(a) Process uncertainty, specifically random temporal variation in recruitment and other population characteristics.
(b) Observation uncertainty, which arises during data collection.
(c) Estimation uncertainty, which refers to the uncertainty about estimated parameter values, as well as about which assumptions are appropriate for the estimation procedures.
(d) Model uncertainty, including functional-form uncertainty, and structural uncertainty (McAllister and Kirchner 2002). The latter term

refers to uncertainty about which model is most suitable for the purpose at hand.

For example, all four of these types of uncertainty arise in the process of stock estimation, and in the determination of biological productivity (McAllister and Kirkwood 1998).

According to the deterministic bioeconomic models of Chapters 1–4, optimal harvest strategies are strongly affected by economic parameters, and this will remain true under biological or economic uncertainty. However, it seems likely that biological (rather than economic) uncertainty will usually dominate fishery management decisions. For example, managing a fish population so as to maintain a fairly large stock level may have both biological and economic benefits, but it is hard to see how uncertainty about future costs or prices would have any major effect on this calculation. Taking account of biological uncertainty will be difficult enough, without adding the dimension of economic uncertainty. (This is not to say that economic circumstances should be entirely left out in the setting of annual quotas, but merely that economic uncertainties can probably be ignored here.)

Finally, the term management uncertainty refers to the extent to which quotas and other regulations may not be complied with. Because of the common-pool nature of the resource, cheating on the regulations may be highly profitable. In some cases scientists have adjusted their TAC recommendations downwards to account for anticipated illegal or unreported catches. Also, IFQ systems have sometimes been criticized for increasing the incentive for misreporting catches. As stressed in Chapter 4, strict monitoring and control of individual catches is essential for the success of any IFQ or ITQ system. Accurate reporting is also essential for scientific purposes.

Two types of risks in fisheries are:

(a) Risks of overfishing, leading to temporary loss of productivity, or to a collapse of the fishery and long-term or permanent loss of productivity. Although overfishing is usually considered to be the main risk in fisheries, one might also include the risk of underfishing, in the sense that the specified TAC is much smaller than is necessary. Although this would have a negative impact on fishermen's current incomes, it would usually also imply increased future stock abundance. Once again, the fishermen's reaction to this risk would depend on the quota allocation system that is in effect. Under (regulated) open access, fishermen would probably object strongly to the low TAC limitation, but under an ITQ system they would recognize that larger future revenues would compensate for short-term losses, thereby enhancing the value of their quotas. Several examples are known in which ITQ fishermen have actually voted for a smaller TAC than recommended by scientists (Grafton et al. 2005).

(b) Risks of loss of ecosystem functioning, including degradation or destruction of habitat, change of ecosystem structure, and loss of biodiversity. In particular, whenever a population is reduced to a fraction of its historical abundance, there is a risk of irreversible system change, which may show up as a "depensation," from which recovery may be slow or non-existent (Jackson et al. 2001).

5.2 Decision Analysis and Fisheries Management

Decision analysis is a sophisticated method for incorporating risk and uncertainty into management decisions. The book *Uncertainty* by Morgan and Henrion (1990) provides a down-to-earth, pragmatic discussion of decision analysis (which the authors call risk and policy analysis), with many examples from environmental, health and other systems. Here I describe the main components of decision analysis, as they might apply to fisheries. My own experience in this area is minimal, but participation in a recent study pertaining to the protection of endangered species convinced me and my collaborators that decision analysis was the best, if not the only viable approach. Although neither the biologists or economists involved in the study had any experience with decision analysis per se, many of its component details were already familiar to us.

The Basic Steps

Here in a nutshell are the main steps taken in a decision analysis. We assume that interest is focused on some marine system, whether a particular fish stock, or an entire marine ecosystem, or some intermediate system. Indeed, the first step is the determination of "boundaries"—exactly what is the problem to be addressed? As for all steps in the analysis, the boundaries may be altered as the project proceeds. However, it is also important to limit the scope and complexity of the problem, since including uncertainty will inevitably result in a complicated analysis in any case.

Keep in mind that the following steps are part of an iterative process. As the analysis proceeds, each previous step may be reviewed and modified if necessary. Of course, eventually the analysis is completed, documented, and reported.

Step 1. Specify the problem.

Step 2. Obtain the necessary data. Identify uncertainties in the data.

Step 3. Identify management options and objectives.

Step 4. Construct a model or models. Identify process, parameter and structural uncertainties in the model.

Step 5. Test the model(s).

Step 6. Run Monte Carlo simulations of the model(s), for each management option.

Step 7. Evaluate the management options.

Step 8. Document the study and obtain peer reviews.

In the fisheries literature, attention has focused mainly on using the methods of decision analysis in stock-assessment procedures and scientific advice for management. For example, McAllister and Kirkwood (1998) list the following steps in a Bayesian framework for evaluating fishery management procedures:

1. Identify each alternative management procedure.
2. Specify the indices of policy performance.
3. Specify the alternative hypotheses.
4. Determine the relative weight of evidence in support of the alternative hypotheses. (This uses the methods of Bayesian inference; see below.)
5. Evaluate each performance index for each management procedure.
6. Present the results to decision makers.

Other articles that employ a similar philosophy include Francis and Shotten (1997), Punt and Hilborn (1997), McAllister and Kirchner (2002), and Harwood and Stokes (2003). A broader approach that also includes economic aspects is discussed by Lane and Stephenson (1998).

Uncertainty is addressed in steps 3–5 of the McAllister–Kirkwood approach. By "alternative hypotheses" is meant a list of alternative models, which may differ in the specification of functional forms within a given model structure (e.g., the stock-recruitment relationship), or in the specification of model structure itself (e.g., spatially lumped versus spatially structured population variables, etc.); see McAllister and Kirchner (2002) for details.

The available data provide a certain degree of support for each alternative model. Specifically, Bayes' theorem implies that the marginal posterior probability for model m_j is given by

$$p(m_j|\text{data}) = \frac{p(\text{data}|m_j)p_0(m_j)}{\Sigma_k p(\text{data}|m_k)p_0(m_k)} \tag{5.1}$$

where

$$p(\text{data}|m_j) = \int p_j(\theta_j)L(\text{data}|\theta_j)\, d\theta_j \tag{5.2}$$

Here $p_0(m_j)$ is the prior probability for model m_j and θ_j is a vector of (uncertain) parameters of model m_j, with posterior distribution $p_j(\theta_j)$. A Bayesian process is also used in calculating these posterior distributions $p_j(\theta_j)$; details are discussed in the cited publication. Also $L(\text{data})|\theta_j)$ is the likelihood function.

The difficult topic of choosing prior distributions is also discussed in the cited references, which also provide technical details about computational algorithms.

For each model m_j the analyst next uses Monte Carlo simulations to find the probability density $F(w_k|m_j)$ for each performance index w_k, given the model m_j. These can be combined into an overall performance index density

$$\hat{F}(w_k) = \Sigma_j F(w_k|m_j)p(m_j|\text{data}) \tag{5.3}$$

Now comes the difficult part—explaining the results to the decision makers! Two methods for presenting the results are (a) graphic, and (b) decision tables. For a graphic presentation one produces graphs of the value distributions $\hat{F}(w_k)$ for each strategy w_k. A decision table displays the same information in tabular form. (Other aspects of presentation are discussed later.)

To consider an example, suppose that the chosen performance index is the total present value of net fleet revenues over a 25-year horizon. Three management strategies are considered, all being reference-point strategies as shown in Fig. 5.1. A logistic model with depensation is used for population dynamics, with one highly uncertain parameter, $B_{\text{crit}} =$ critical stock level, which has the property that a crash occurs (with no subsequent recovery) if $B(t)$ ever falls below B_{crit}. The probability distribution for B_{crit} is obtained from meta-analysis of other similar fisheries. Many other uncertain parameters (e.g., the biological parameters r and K) or stochastic processes (recruitment) could be included in the model, if desired.

Note that the three strategies of Fig. 5.1 trade off yield and precaution

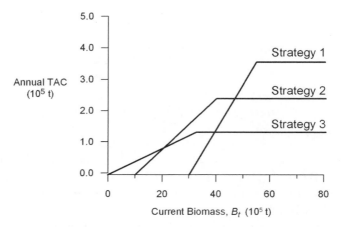

Figure 5.1 Three reference-point harvest strategies.

Table 5.2. *Distribution of 25-year economic yield*
$\hat{F}(w_k)$ *for the three reference-point strategies shown*
in Fig. 5.1.

Value ($ × 10^8)	$k = 1$	$k = 2$	$k = 3$
5	.06	0	.10
10	.10	.10	.15
15	.17	.30	.20
20	.20	.35	.25
25	.21	.20	.20
30	.16	.05	.10
35	.08	0	0
40	.02	0	0

in different ways. Strategy 1 allows high catches from a healthy stock, but
cuts off harvesting at the first sign of overfishing. Strategy 3 maintains
low catches, but never completely shuts down the fishery. Strategy 2 is
intermediate.

Table 5.2 is a decision table for the three strategies w_k. The same
information is shown graphically in Fig. 5.2. Suppose that these results
are presented to the decision makers. Which is the preferred strategy?
How can this be decided?

One approach is to calculate the expected ("average," or "best estimate") value for each strategy. These expected values, $215, 190 and
180 million respectively, are indicated by the arrows in Fig. 5.2. This

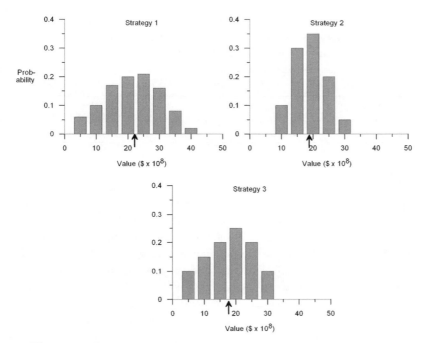

Figure 5.2 Graphical representation of the information given in Table 5.2.

averaging procedure would identify Strategy 1 as the best strategy. According to Punt and Hilborn (1997), "We have found that managers gravitate toward the simplest presentation and that the expected value of consequences is often all that they want to see." Clearly, however, this averaging rule makes a mockery of decision analysis. Basing decisions solely on expected values (i.e., averages) simply *ignores all aspects of uncertainty and risk.* Although these uncertainties and risks are identified during the decision analysis, in the end they play no role in the decisions. (Taking averages at the end of the analysis can give different results from using averaged parameter values throughout, but this is not the point here.)

In Fig. 5.2, note that Strategy 2 is notably less risky than Strategy 1: the distribution $\hat{F}(w_2)$ is more sharply peaked, and less widely spread out, than $\hat{F}(w_1)$. In other words, the future benefits from Strategy 2 are more predictable than are those of Strategy 1. (Strategy 3 is less profitable than Strategies 1 and 2, and at least as risky as either.) A bona fide decision analysis would surely consider Strategy 2 as a strong candi-

Table 5.3. *Expected revenue and probability of collapse for two reference-point strategies shown in Fig. 5.1, as well as a fourth strategy.*

	Expected Revenue	Probability of Collapse
Strategy 1	$215 million	.182
Strategy 2	190 ″	.137
Strategy 4	175 ″	.045

date. Indeed, such an analysis would explicitly recognize risk reduction as part of the decision criterion in the first place.

Here perhaps we can learn a lesson by thinking about portfolio analysis. An investment advisor will usually ask the client "What are your financial objectives?" What the advisor is trying to elicit is the client's attitude towards risk. After careful questioning, the advisor may come up with a portfolio strategy, such as 50% growth, 30% income and 20% highly speculative. If the client agrees with this, the portfolio of investments will be selected to follow those guidelines. Most financial advisors explain that intelligent investing requires the investor to decide what his or her attitude toward risk is.

Exactly the same principle holds—or should hold—in fishery management. Risk management requires not only risk assessment (e.g., in stock assessments), but also specific determination of the decision maker's degree of risk acceptance and avoidance. Basing management decisions on expected values alone by-passes risk management entirely. The quotations listed earlier suggest that few fisheries scientists yet recognize this basic principle of policy analysis and decision making.

Returning to Table 5.2, note that, for example, Strategy 1 has a 16% probability that the 25-year discounted revenues will be less than or equal to $100 million, whereas this is reduced to 10% under Strategy 2. The managers might decide to sacrifice some of the expected revenues in favor of greater certainty. The foregone expected revenue ($15 million, or 7% of the expected revenues) would be considered a "premium" paid for the lowered risk.

The managers might also want to know why Strategy 1 is more unpredictable than Strategy 2. For example, perhaps Strategy 1 is more likely to cause a collapse of the fishery. Monte Carlo simulation can readily generate this prediction; see Table 5.3.

Here Strategy 4 is similar to Strategy 1 but with a lower maximum

TAC. Note that Strategy 4 seems to be much less risky than 1 or 2. This extra information might be considered critical. For example, the decision makers may have adopted a precautionary approach, such as maximizing 25-year expected (discounted) economic yield, subject to maintaining at least a 90% estimated probability that the population will survive over the 25-year period.

But is there some way to "scientifically" determine the best strategy? Recall the example of investment portfolios—optimal investment involves both expected yields and acceptable risk levels. Surely most people would reject a management philosophy that put an important fishery resource at almost a 20% risk of collapse over a 25-year time span. Perhaps the central government would mandate that all marine fisheries be managed to ensure a 25-year probability of survival of at least 90%, according to the best scientific evidence. This would also ensure that proper risk (decision) analysis is carried out for each fishery. See Chapter 7 for further discussion.

This example raises yet another issue. In the illustration we have imagined that the only policy choice is which reference-point harvest strategy w_k is to be used. In reality many other management options may be available, such as,

- closed areas (marine reserves)
- protection of spawning concentrations
- maintenance of a broad age-structure
- control of by-catches from other fisheries

The list is almost endless, even without including the possibility of using individual quotas. In principle each of these options, or any combination thereof, could be subjected to a decision analysis. Only the time and expense of such studies would limit their use. With the experience gained from a first decision analysis, perhaps other scenarios could be analyzed fairly routinely.

This artificial illustration does indicate that there may be wide scope for decision analysis to be used on a regular basis in fishery management. It seems likely that management will continue to evolve in two directions simultaneously, decision analysis on the one hand, and pseudo property rights (ITQs, etc.) on the other. At the same time, and perhaps supporting these trends, ecosystem considerations may begin to play a greater role, at least in terms of directing conceptions about marine systems (Walters and Martell 2004; Mangel and Levin 2005). It seems likely, however, that single-species models will continue to be used for some time to come (Quinn and Collie 2005).

Residual Uncertainty

The above illustration considers only a single model of population dynamics (but with an uncertain parameter). What if there is also uncertainty about which type of model to use? Eq. (5.2) shows how to evaluate the relative believability for each of n different models m_j. Also, Eq. (5.3) shows how to calculate the overall payoff distributions $\hat{F}(w_k)$ for the different management strategies w_k, for any given model.

Table 5.4. *Expected 25-year discounted revenues and collapse probabilities from a decision analysis using two competing models.*

	Expected Revenue			Probability of Collapse		
	m_1	m_2	Overall	m_1	m_2	Overall
Strategy 1	300	250	290	.19	.11	.174
Strategy 2	250	150	230	.04	.33	.098

Table 5.4 shows model predictions for the case of two competing models, and two management strategies (as before, the data are entirely artificial). The estimated posterior probabilities of models m_1 and m_2 are assumed to be 0.8 and 0.2 respectively. Looking just at the overall predictions, one might prefer Strategy 2 because of its smaller risk of collapse. But note that Strategy 2 is in fact highly risky if it happens that model 2 is the correct one. Why this reversal?

The overall probabilities of collapse are given by

$$p(\text{collapse}|\text{data}) = \Sigma_j p(\text{collapse}|m_j, \text{data})p(m_j|\text{data}) \qquad (5.4)$$

In other words, this is the weighted average from the separate models (as in Eq. 5.3). Because model 2 has a low posterior probability, averaging conceals the risk associated with model 2, which may nevertheless be the correct model. As we saw earlier, taking averages inevitably hides the risks involved (recall the "best-estimate" discussion).

A real-world example of this process may heave been instrumental in the collapse of the Northern cod fishery of Newfoundland in 1991 (Hilborn et al. 1993). Scientists worked with two stock-assessment models, one based on randomized sample surveys and the other on CPUE data. The results being in disagreement, an average was taken to obtain an estimate of current stock levels. The resulting TAC would have been safe if the CPUE-based estimate, or the averaged estimate was correct, but unfortunately the survey results were in fact correct and the CPUE results incorrect. A proper decision analysis would perhaps have forced decision makers to address this discrepancy directly.

As far as I am aware, no actual decision analysis in any fishery has ever gone into this much detail, although doing so is clearly possible with present-day techniques. My guess is that many, if not most fisheries actually involve just this kind of model uncertainty. The question is, how should decision makers respond to a situation in which competing models indicate very different management strategies? Neither model can be rejected outright.

Two possible answers are:

1. Perform research designed to clearly identify the correct model. This is known as "adaptive management," although that phrase also has other implications (Walters 1986).
2. Consider alternative management strategies that are likely to succeed regardless of which model is correct. This is called "robust management." This term is defined by Charles (2001) as management that is "reasonably successful in meeting societal objectives, even if (a) our current understanding of the fishery (notably the status of the resource), its environment and the processes of change over time turns out to be incorrect, and/or (b) the actual capability to control fishing activity is highly imperfect."

Charles (2001) also states that "The move to robust management requires a rethinking of the philosophy of management, including the adoption of new structural and decision-making tools, notably the precautionary approach and the ecosystem approach." I would add to this list the decision-analysis approach, which has the potential of making the precautionary approach operational, and the ITQ approach, which can encourage fishermen to adopt the principles of risk management.

What sort of management strategies would be satisfactorily robust? Certainly the strategy of setting a single TAC for a large fish stock seems fraught with risk, and this has been demonstrated by historical experience. A more robust strategy would include diversification as a primary feature. For example, the total quota might be divided into area-specific sub-quotas, high in some areas and low or zero in other areas. Then errors in stock assessment, or in controlling the catch, would be less likely to lead to a collapse of the fishery. Many studies of marine reserves, for example, have stressed the precautionary, or bet-hedging nature of a strategy that protects a portion of the stock; see Sec. 5.3 for further discussion.

Several authors have suggested that effort quotas are more robust against various types of uncertainty than are catch quotas. First, a specified level of effort is less likely to devastate a stock that is smaller than the estimated size. Second, effort may be more accurately controllable than are catches. Whether these presumed advantages are real or illusory depends on whether nominal effort (which is usually what is controlled) is in fact proportional to fishing mortality. In other words, in preferring effort quotas over catch quotas, there is a tacit assumption that the Schaefer catch equation $h = qEx$ (q = constant) is more or less valid, when E equals nominal (controlled) effort. This point was discussed in Chapter 2, and need not be elaborated here. It is worth noting, however, that many discussions of the desirability of effort quotas fail to define the term "effort." Do these authors hope that their readers will not realize that there are two different concepts of fishing effort, with contrasting operational implications?

Another suggestion is to combine catch and effort controls, but this strikes me as just stacking on more regulations in the hope that this will fix the game of competitive fishing. Multiple controls should not be necessary in an IFQ system; experience shows that the fishermen themselves tend to optimize effort in this situation. In general it seems likely that robust, risk-limiting harvest strategies would often emerge deliberately in IFQ/ITQ systems. It is not so clear what would happen in an IEQ system, however, since in this case the fishermen will be motivated to "get the most" out of their effort quotas.

To summarize, it seems unlikely that the participants in a free-for-all TAC-based fishery would favor a decision-analytic approach to management, particularly if this results in a reduced-quota, precautionary management regime. As pointed out earlier in this book, under such circumstances fishermen have little or no incentive for resource conservation or enhancement. Support for risk management and decision analysis might well be forthcoming, however, under ITQs, because controlling risks should enhance the value of a quota. Time will tell if ITQ fisheries automatically move towards risk-limiting harvest strategies, or if additional government control of fishing strategies is required.

5.3 Marine Reserves as Insurance

In 1992 I had a student, Tim Lauck, who was interested in marine reserves. After a careful literature search, Tim came up with a handful—maybe three to five—published papers on the topic. Marine reserves were a dead issue.

Before writing this sentence in 2005, I ran an Internet search for "Marine Reserves." I got 153,000 hits. Marine reserves are a hot topic.

Why the change? It seems that people suddenly got wind of the crisis in marine fisheries in the early 1990s. Overfishing, spectacular fishery collapses, the gradual death of coral reefs—these are big news stories. Fish stocks within the recently established 200-mile zones have been badly overfished. Bottom trawling has destroyed benthic communities in many supposedly carefully managed areas. If this is the best that our governments' departments of fisheries can do, something has to change. It was only natural to raise the idea of protecting a bit of what's left by setting aside fully protected marine reserves (sometimes called marine protected areas, or MPAs).

Experience with existing marine reserves has demonstrated that fish communities can be reconstituted, sometimes quite rapidly, when fishing pressure is removed. Maybe marine reserves would actually help the fishermen by providing a secure source of supply to otherwise depleted areas. How about adding marine reserves to the list of management strategies—or even using them exclusively? Let the fishermen operate freely outside the reserves, but exclude them completely from the reserves. The latter is easily accomplished if all vessels are required to carry satellite transponders.

Come the modelers! It turns out that it won't work, except under special circumstances (Holland and Brazee 1996; Hannesson 1998; Hastings and Botsford 1999). The fishermen excluded from the reserve just intensify their effort outside, especially barely outside, the reserve. In most cases the economic benefits will be nil, if not actually negative.

But don't these studies overlook something important (an inspection of the papers will reveal how restrictive the assumptions are)? The basic idea of reserves is protection against management limitations and errors (Lauck et al. 1998). We know that, with competitive fishing, the fishermen will always search out the best available fishing areas, reducing the stock in each area to the local bionomic equilibrium—unless a TAC closes the season before they can achieve this. The fishermen—with or without a TAC—will have zero incentive to protect habitat, or to leave

large fish in the sea as spawners. A marine reserve would do both—
protect habitat and spawners. According to Roberts et al. (2005), "In
addition to recovering stocks of target species, other key fishery management benefits claimed for marine reserves include the development
of natural age structures of exploited species, protection of genetic variability, restoration of ecosystem integrity, more predictable and often
higher catches *and insurance against management failure.*" (emphasis
added). These authors go on to assert that "For example, the tool of
choice for managing fisheries in Europe, total allowable catches and national quotas, has the least conservation value of any management tool
available ... and has failed to deliver sustainable fisheries in the past
.... To achieve sustainable fisheries and protect non-target species and
their habitats, fishery management must embrace tools that include prohibition of the most damaging gears, areas closed to particular gears,
precautionary quotas, bycatch quotas, and modification of fishing gears
and practices to reduce the collateral damage of fishing. Such measures
will not, in themselves, be enough without the widespread introduction
of fully protected marine reserves."

In short, marine reserves serve many purposes. For the present discussion, their main role is as an insurance policy against uncertainty
and management failure. Thus the possibility of using reserves should
be included in the decision analysis of a fishery. In any specific case,
the design of reserves itself would also be facilitated by decision analysis, given that much of the necessary information is likely to be quite
uncertain. Some articles that discuss reserve design are: Murawski et
al. (2000); Sanchirico and Wilen 2001; Apostalaki et al. 2002; Roberts
et al. 2005. The volume edited by Sumaila and Alder (2001) discusses
economic aspects of marine reserves.

5.4 Summary of Chapter 5

This chapter began by discussing the minor role assigned to uncertainty
in formulating fisheries management advice, until quite recently. At
present, however, Bayesian methods are beginning to be used to estimate
uncertainty, for example in stock assessments. Still, the presentation of
uncertain information to decision makers has remained problematic—
what are the decision makers to do with such information?

A common reaction is to consider only expected values, for example long-term average catches, or revenues. This procedure completely

negates the purpose of studying uncertainty in the first place. It is akin to asking your stock broker to choose your stocks on the basis of expected returns only, giving no consideration to risk. Responsible brokers do not manage portfolios in this fashion.

How, then, should fishery managers take risk into account? This question is addressed in Sec. 5.2, under the topic of decision analysis. For example, the risk of severe overfishing (leading to possible collapse) can be estimated for different management strategies, and only those strategies that meet some pre-specified precautionary criterion will be considered acceptable. The supporting risk analysis will consider all identifiable sources of uncertainty, taking cues from the study of historical successes and failures elsewhere (Smith and Link 2005).

Section 5.1 itemizes kinds of uncertainty and risk in fisheries. Biological uncertainties include process uncertainty, (i.e., random fluctuations in recruitment and other characteristics), and informational uncertainty, (i.e., incomplete knowledge about the relevant biological system). Informational uncertainty includes both parameter and structural uncertainty in the specification of models—for example, whether spatial structure needs to be explicitly represented in a stock-assessment or management model of a given population. Bayesian methods for handling informational uncertainty are discussed in Sec. 5.2.

Risks inherent in fisheries include, first, the risks of inadvertent overfishing, which in extreme instances can lead to the collapse and nonrecovery of the resource, and second, risks of causing ecosystem disfunction of various kinds. Risk assessment activities would try to identify these risks, and their severity under alternative harvesting strategies. Unfortunately, work has barely begun on developing protocols for risk assessment in fisheries.

In Section 5.2 we described the methodology of decision analysis. This technique consists of an iteratively modified sequence of steps that (i) specifies the problem, (ii) identifies and quantifies the uncertainties, (iii) specifies the management options and objectives, (iv) builds and tests models (perhaps including alternative models), and then (v) runs model simulations to produce posterior distributions of objective values. This procedure facilitates decisions that are consistent with management objectives, while recognizing the implications of uncertainty.

Decision analytic methods have recently begun to be used in the areas of stock assessment and model evaluation, but not yet in the broad sense of determining management objectives and basing decisions on risk as well as on expected benefits. As argued in Chapter 7, moving towards fully professional management of marine fisheries, with the dual objectives of sustainability and profitability, will probably require the comprehensive re-design of current management institutions and techniques.

The penultimate section of Chapter 5 discusses one management tool, marine reserves, that seems destined to become a major component of fishery management. Marine reserves explicitly recognize the imperfectability of decision making in the highly uncertain world of the marine environment. They are the epitome of precautionary management for conservation and exploitation of our ocean heritage.

An important question is whether existing management structures are capable of using decision analysis. Chapter 7 discusses the need for developing more broadly based decision-making institutions in fisheries.

6

Case Studies

This chapter briefly describes five fisheries in which ITQs or related systems have recently been instituted. (See Ocean Studies Board 1999 for other case studies of ITQ fisheries.) The five fisheries cover a variety of species and fishing techniques. In most cases fish stocks were severely depleted, or in imminent danger of becoming so, when the ITQ system was initiated. In all cases the new management regime has apparently stemmed or reversed the depletion of the stock.

Economically speaking, the results are less clear. Vessel owners are notably reluctant to provide details about the profitability of their operations, and few of the cited references include detailed cost data or profit estimates. Profitability is sometimes inferred from the fact that quota prices have increased over time. An alternative explanation is that the price increase is mere speculation. Without actual financial data the two explanations cannot be separated.

6.1 Georges Bank Sea Scallop Fishery

The Atlantic sea scallop is distributed over the Northwest Atlantic continental shelf, from the Gulf of St Lawrence to Cape Hatteras, with a major population on Georges bank off New England. Scallops, caught by bottom dredging, are processed at sea, the product being fresh or frozen scallop "meats" typically sold in stores and restaurants. The description given here is condensed from Repetto (2001) and Edwards (2002). Repetto compares the management system for US and Canadian fisheries, calling this a "natural experiment in fisheries management." Roughly speaking, the American fishery is currently at regulated bionomic equilibrium, with near-zero economic rents, while the Canadian fishery, based on a form of individual quotas called Enterprise Allocations (EAs), generates large (but unspecified) rents for the industry.

Sea scallops have been harvested commercially on Georges Bank since the 1880s, initially with small sailing vessels using hand-drawn dredges, but currently with large (over 150 ton) high-powered vessels using advanced fishing and navigational technology. Vigorous competition between US and Canadian vessels, both before and after the 1977 declarations of 200-mile zones, was halted in 1984 when the International Court in The Hague set the limit separating the EEZs of the two countries, cutting across the Georges Bank grounds. Since that date, the American fishery has been under the management of the North East Fisheries Management Council, and the Canadian fishery under Canada's Department of Fisheries and Oceans. As will be seen, the two management regimes have developed very differently, providing a case study for the effectiveness of individual-quota-based management systems.

The American Fishery

Annual landings in the US fishery peaked in 1990–1991 at 38 million pounds, well in excess of the estimated maximum sustainable yield of around 20 million pounds. Prior to 1994 the only regulation imposed on the fishery was a minimum size-limit (50 meats per pound), introduced in 1982. This regulation, intended to reduce the catch of smaller scallops, had the unexpected consequence of causing fishermen to soak small scallops in ice water, thereby increasing their weight but decreasing their value due to loss of flavor. Between 1977 and 1993 entry of full-time scallop vessels increased eight-fold, an expansion that ceased with the introduction of limited licensing in 1994. Each of the 357 licensed vessels was issued an annual effort quota, in terms of number of days at sea. Initially as high as 200 days per year, effort quotas were reduced to 120 days by year 2000, 51 days in 2001 and 34 days in 2004 as part of a ten-year stock rebuilding program. These quotas were insufficient to cover annual fixed costs for most vessels. However, effort quotas could not be consolidated, owing to a prohibition on transferring quotas. This prohibition was upheld in December 1997 by the North-East Fisheries Management Council. Edwards (2002) estimated that allowing multi-permit companies to transfer effort quotas between vessels would save about $200,000/year in fixed costs for each unused vessel. Exactly why some fishermen groups have opposed transferability is unclear; Edwards suggests that permit transfers would favor some groups over others. Estimated average profit per vessel day in the Georges Bank scallop fishery declined from \$2,900 in 1977 to −\$400 in 1998.

Another important development in this fishery was the establishment of closed areas, amounting to over 30% of the Georges Bank site, beginning in 1994. These areas were closed to both ground fishing and scallop dredging, for purposes of stock recovery. Temporary, controlled access to the unprecedentedly high scallop biomass that had built up on Closed Area II was allowed in 1999 and 2000, but this has not been continued in subsequent years. One plan under consideration calls for the use of rotating closed areas, which would allow scallops to grow to larger sizes before being harvested. Scallops recruit into the fishery at about age 3, and thereafter increase in size by a factor of four by age 5. Catching young scallops thus severely reduces sustained biomass yield in the fishery (and further impacts revenues, since price per pound increases with size).

The question arises whether some alternative management system for the American Georges Bank scallop fishery could lead to substantially increased incomes for all those involved in the fishery. Experience with the Canadian fishery, as described below, strongly suggests that this is correct. Repetto (2001, p. 263) muses that "over the coming years the US scallop fishery will move toward and finally adopt a rights-based regime, putting itself in a position to realize some of the economic benefits that the Canadian industry has enjoyed over the past decade."

The Canadian Fishery

Since 1985 the Canadian sea scallop fishery has operated under an Enterprise Allocation (EA) system, wherein each fishing company owns an annual percentage share of the total annual catch (TAC). At the government's insistence, initial allocations were determined by negotiations among the nine companies themselves. Subsequent quota transfers are permitted, subject to various regulations, but only with government approval. How many vessels a company uses to catch its annual quota is left entirely to the company.

A government research program, financed by the industry, regularly surveys the scallop population and provides information pertaining to harvest potential. The industry currently prefers a conservative TAC, in order to stabilize the annual harvests in the presence of recruitment fluctuations. These surveys map bottom topography and provide information on local density of scallops, thereby improving the efficiency of dredging operations, while also reducing the bycatch of groundfish.

The annual mortality rate in the Canadian scallop fishery under EAs has settled at about 20% at ages 4–7 yrs, and virtually zero for age 3 scallops. These mortality rates have resulted in a high standing biomass, with high yields of the more valued large scallops. The large stock provides for high catch per day and also hedges against occasional years of low recruitment, permitting relatively stable annual harvests. To reiterate, these conservation-oriented strategies are strongly supported by the scallop industry.

Assessment

Regarding public equity, the Canadian government began collecting royalties, in January 1996, of $547.50 per tonne of quota. Repetto (2001) does not provide annual catch data, but Canadian catches on Georges Bank are smaller than those of the American fleet, and perhaps amount to some 3–5 million pounds/year. This would imply an annual government royalty revenue of about $1 million per year. Though not negligible, this seems unreasonably low, assuming that ex-vessel prices of scallop meats probably exceed $20,000/tonne, implying an estimated gross annual industry revenue of about $40 million. (Both Repetto and Edwards avoid any mention of actual price or cost data for the scallop fishery. As Repetto puts it, "It is notoriously difficult to persuade fishermen to report or to talk about how much they're making ... .)"

Why are the American and Canadian scallop fisheries so different in terms of economic performance? How did the American fishing industry succeed in lobbying Congress to outlaw ITQs (with a few specific exceptions), which are the best known method of protecting fishery resource biologically and economically? To quote Repetto again, "If fisheries are to be managed in the public interest, what party should be held accountable for this waste of resources and sacrifice of potential economic benefits, the Congress, NMFS [the National Marine Fisheries Service], or the industry?"

For the case of the US sea scallop fishery, part of the answer may lie in the heterogeneity of that system, compared to the Canadian case. According to Edwards, one group of advisors recommended against using ITQs in this fishery on the basis that it had taken 13 years to negotiate quota allocations in the Atlantic surf clam fishery. Other groups have favored ITQs.

The title of Edwards' paper, "Rent-seeking and property rights formation in the U.S. Atlantic sea scallop fishery," perhaps provides the clue to this situation, and to similar situations in other fisheries located within national EEZs elsewhere. Introducing a rights-based system (for example, ITQs or community rights) into a fishery will in fact generate positive, and potentially large future rents. Even at the planning stage, rent-seeking behavior will therefore emerge. Since most of the future rents will be captured by the initial rights recipients, it is only to be expected that arguments over who is awarded these rights will be vigorous and protracted. For example, even though ITQs were finally allowed in the 1992 revision of the Magnusson-Stevens Act, few ITQ systems have yet been successfully negotiated and launched.

In some cases governments have settled the initial ITQ allocation issue by simply awarding quotas to specific individuals. Examples in Canada include the Atlantic sea scallop fishery, and the British Columbia herring-roe fishery.

Is there some better method for allocating fishery rights to specific individuals? This question is addressed in Sec. 4.8, where it is suggested that the retention of a substantial portion of resource rents by governments can achieve two important purposes:

1. Social equity is preserved by allocating a substantial portion of resource rents to the resource owner, the public;
2. Rent-seeking behavior is reduced when excess rents accrue to the government.

6.2 The Italian Adriatic Clam Fishery

Beginning in the 1970s a lucrative fishery for clams *Chamelia gallina* developed along a 750-km stretch of the Adriatic coast of Italy (Spagnolo 2004). Fishing is carried out by hydraulic dredges, standardized vessels (approximately 10 tonnes) that dredge up a mixture of clams and sediment, which are then separated using a hydraulic process.

Entry into the clam fishery was controlled by a government licensing system, with 384 licenses issued by 1974, a figure that had increased to over 800 by 1994. Early landings exceeded 100,000 tonnes/year, but by 1993 this had declined to 38,000 tonnes/year. In addition, the average size of clams had fallen to such an extent that the price declined from € .50–.80 to .24 per kg. Although a wide variety of input and output controls (e.g., limited vessel size, daily catch, fishing time and area) had been instituted, many of these were widely violated owing to enforcement difficulties and the large benefits obtainable from non-compliance.

In response to this situation, the Italian government in 1996 introduced a new management program, hereafter referred to as "Plan I," with the following characteristics:

First, 12 management "Consortia" were established, each covering a specified section of the fishing grounds. Only fishermen who registered with a given Consortium were eligible to benefit from financial resources allotted to the Consortium. Members were granted the right to fish within the Consortium area. In fact 100% of licensed clam fishermen joined a Consortium.

Second, the Consortia were given extensive powers to manage the clam fishery in their areas. Consortium members were admonished to cooperate in developing and implementing appropriate management strategies for their area.

Third, transferability of licenses was allowed, but only if accompanied by transfer of vessel ownership. This regulation successfully prevented concentration of licenses.

Fourth, a voluntary vessel buy-back program resulted in the removal of 36 dredges (4.5% of the fleet), at a cost of € 130,000 per vessel, calculated on the basis of the market value of a license.

Finally, subsidies were provided for restocking the clam population.

The objectives of the Plan were to establish decentralized management of the fishery, to transfer management responsibility to the fishermen, and to initiate a form of property rights in the fishery. These rights were assigned to each of the Consortia, not to individual fishermen.

In 1998 a revised plan ("Plan II"), recommended by Consortia members, was introduced, with the aim of removing additional vessels in areas where stocks were still depressed, and also of increasing the level of management responsibility of the Consortia. As a result, 109 additional dredges were removed permanently from the fishery (at € 193,000 per vessel), and a smaller number were subsidized for temporary withdrawal. The total costs of both Plans (borne by the government) were as follows (figures are in € millions):

Purpose	Plan I	Plan II	Total
Permanent withdrawal	5	21	26
Temporary withdrawal	–	10	10
Restocking, etc.	22	5	27
TOTALS	27	36	63

Fishermen were informed that no further funds would be forthcoming beyond Plan II.

Total annual production of clams, and the price obtained, for the period 1998-2002 were:

Year	Annual Production (tonnes)	Price (€/kg)	Gross Value (million €)
1998	25,000	1.80	45.0
1999	38,000	1.10	41.8
2000	30,000	2.20	66.0
2001	27,000	2.90	78.3
2002	13,000	4.50	58.5

These data indicate a strong supply-price relationship (but the average size of clams also affects price), suggesting the desirability of an industry-wide production and marketing strategy. Such a strategy, initiated in 1999 via coordination of all Consortia, has "brought the profitability of producers to a record level, while reducing the level of exploitation" of the clam resource (*ibid.*). Annual catches are controlled by a combination of fleet reduction, season length and daily catch limits per vessel. (Note in passing that the latter two controls essentially imply a fixed catch quota per vessel, even if this is not the intent.)

As a result of these developments, the value of licenses has risen dramatically:

	1996	2002
Gross annual profit per vessel	€ 30,000	€ 77,000
License value	€ 130,000	€ 500,000
Imputed interest rate	23%	15%

The decline in the imputed interest rate (obtained by dividing gross annual profit by license value: $i = R/PV$) suggests that vessel owners in 2002 had greater confidence in future revenues than they did in 1996.

Assessment

As a result of the decentralization program that transferred management responsibility to the 12 Consortia, the Adriatic clam fishery has been transformed from an inefficiently managed and overexploited fishery to one based on community rights. Controls on vessel size, season length, and daily catch limits seem now to be accepted, and have apparently been sufficient to generate near-optimal rents. Profitability having been thereby increased, license values have risen almost four-fold. Indeed, the overall increase in license values from 1996 to 2002 was roughly € 250 million, compared to the government's initial investment of € 63 million for setting up the new Plans. Whether the windfall gain should fairly accrue entirely to the fishing industry is a question that is not addressed in the cited reference. As Spagnolo says, the system is "highly profitable for those who have held the license from the very beginning," although any newcomers who buy up an existing license can only expect standard returns on their investment.

Although the new management program appears successful, Spagnolo (2004) points out that various environmental crises (anoxia, predators, parasites, toxic algae) can sometimes severely impact the clam population. This could motivate fishermen to overexploit areas not affected by such events, "so destroying all the achievements gained so far." Perhaps an industry-financed insurance program is needed to respond to such risks.

Finally, why has the new management system been so successful? Why couldn't similar results have been achieved through the original national management strategy?

We can only provide speculative answers to these questions. First, it seems that the clam fishermen perceived the central authority as the "enemy," interfering with their fishery. But when catches, and the size of clams in the catch, declined sharply in the 1990s, the fishermen were ready for a new approach. Designed with the purpose of localizing decision making, the system of 12 Consortia explicitly transferred management responsibility to these local community institutions. This change seems to have generated a strong incentive for cooperation, to ensure that the profitability of the fishery was retained and enhanced. The sedentary nature of the resource, together with the homogeneity of the fishing fleets, contributed to the tractability of the new management system. Plan II, a fine-tuning of the first Plan, ensured the sustained profitability of the fishery.

6.3 Iceland's ITQ Fisheries

Iceland's current system of individual transferable quotas, or ITQs, which covers all of its major fisheries, was introduced in a series of separate steps, in response to crises in the various stocks (Arnason 1993). First, immediately after the declaration of its 200-mile zone in 1976, the government introduced non-transferable IQs into the herring fishery, which had been under a full moratorium since the collapse of herring stocks in the late 1960s. The individual quotas were, however, very small, and with support from the industry, transferability was allowed starting in 1979. All quotas were required to be associated with licensed vessels, a rule that prevented the accumulation of large quota holdings by any individual.

Individual quotas were next introduced into the capelin fishery, in 1980, again in response to a severe decline of the stocks. By 1988 the capelin quotas had also become transferable.

Following the 1976 extension of fishing zones, Iceland's demersal fisheries (cod, haddock, saithe, redfish and other species) were managed using a TAC, or total annual catch quota, system. Entry to the demersal fisheries was uncontrolled, with the result that capacity gradually increased and the season length reduced. A sharp decline in demersal stocks led, in 1994, to the introduction of ITQs, although an alternative option, individual effort quotas, was also allowed. The effort-quota option was widely used, and more than half of the demersal catch was taken under effort quotas, from 1986–89.

In 1990 a uniform ITQ system was adopted for all of Iceland's fisheries; effort quotas were abandoned at that time. According to Arnason (1993, p. 207), the final acceptance of the ITQ system by the industry "must be attributed to the potentially immense economic benefits of the vessel quota systems that have now become apparent to most of the participants in the fishery." As discussed later, not all Icelanders seem to agree with this assessment.

The principal features of the 1990 ITQ system were: (a) quotas were specified as shares of the annual TAC for each species, as set by the government on the basis of recommendations from fishery biologists; (b) vessel quotas were permanent, infinitely divisible and transferable to other licensed vessels; and (c) quota fees, intended to defray the costs of monitoring and enforcement were limited to 0.2% of estimated catch value. Although fees have increased somewhat, by the end of the decade, quota fees only covered approximately 50% of management costs (Eggertson 2004).

Two questions that arise are, first, whether the ITQ system has proven to be successful in meeting its objectives, and second, whether the people of Iceland consider the system to be reasonable and fair. Regarding the second issue, Eggertson (2004) asserts that "Since Iceland gained its independence in 1944, few domestic issues have caused such intense and widespread anger in many quarters as the 'free quotas'." To be fair, however, few people apparently thought much about the long-term implications of the ITQ system, with minimal quota fees during its gradual development in response to crisis situations.

To begin, in purely biological terms it does seem that the present ITQ system has at least succeeded in preventing the collapse of vulnerable stocks of both pelagic and demersal species. The herring and capelin fisheries are sustainably productive (although capelin stocks experience wide natural fluctuations in abundance). Less success has been registered for demersal stocks. For example, harvestable cod stocks were estimated at 1.5 million tonnes in 1980, 0.5 million tonnes in 1992, and 0.75 million tonnes in 2000. Whether this recovery failure results from continued overfishing (government TACs often exceed the scientists' recommendations, and catches may have exceeded the TACs) or other factors is unknown. Neither cited reference mentions the possibility that habitat degeneration from trawling activities may have decreased the productivity of the fishing grounds; see Kaiser et al. (2002).

As far as economic rents are concerned, it appears that ITQs have resulted in positive rents in the herring and capelin fisheries. The data needed to decide the issue for demersal fisheries are not available. Arnason (1993) says that "various indicators including quota values in the demersal fisheries strongly suggest that significant rents are being generated by the system." However, Eggertsson (2004) reports that "Little or no progress has been made with stock rebuilding in the demersal fisheries, and recent government commissioned reports find virtually no evidence of resource rent accumulating in the industry... Yet both rental prices and purchase prices of individual quotas have increased continuously in recent years and reached very high values."

The high rental price for quotas seems particularly bizarre—why would a vessel owner pay such a price unless he expects to quickly recoup his investment with a good catch? Perhaps exceptionally high catch rates become available at certain times or locations, and fishermen are willing to pay dearly to be able to exploit such opportunities. No ITQ or other management system would be able to react to micro-events of this kind; high short-term rental prices may be inevitable under such circumstances. As in other situations, fair catch taxes would ensure that at least some of the windfall gains would accrue to the public purse.

Assessment

Eggertsson (2004) delineates several additional shortcomings of Iceland's ITQ system. He asserts, for example, that theoretical models in fisheries management have treated the problems of "moving a fishing industry toward optimal utilization of the resources ... in a casual manner, ignoring monitoring and enforcement, perverse incentives, political economy, and incomplete knowledge." As a result, "economics created false optimism about the task of managing ocean fisheries." ITQs were introduced in Iceland without proper considerations of these questions. In addition, "The system of ITQs in Iceland does not imprint the incentives of proprietorship on the industry, partly because attempts have not been made to create such incentives. The government itself has assumed the owner's traditional maintenance and monitoring roles [in managing the fisheries] .... Research indicates, however, that internal governance is most effective when the rules emerge from within the user group rather than being imposed by a distant political authority (Ostrom 1990)."

The main thrust of Eggertsson's criticisms, however, concerns the controversial equity aspects of the ITQ system: "Few people were conscious of the potential values of the quota rights that the 1984 law transferred to the fishing industry ... In Iceland the potential value of the rent from ocean fisheries is large relative to GDP. Many members of the public feel that they have been robbed of their share in the rent of the national commons."

Fortunately, Iceland's ITQ system is subject to regular restructuring. It seems likely that many of its faults will eventually be corrected.

6.4 Namibia

The Southwest African country of Namibia attained independence in 1990. Prior to that time, uncontrolled fishing by foreign fleets had resulted in the near collapse of many stocks of fish in the rich Benguela upwelling system off Namibia. Since independence, however, a new management regime has been put into place, characterized by conservative

TACs, rights-based fishing activities, and strong control and surveillance. As a result, the depleted stocks have recovered and Namibia's fisheries are now providing substantial yields, both biological and economic (Nichols 2004).

The country's newly elected government, recognizing the potential for a major contribution to the Namibian economy from development of its fisheries sector, quickly established the Ministry of Fisheries and Marine Resources, which was given the responsibility of rebuilding the fisheries and setting up a management system designed to obtain sustainable economic rents from the resource, and also to ensure that these benefits accrued to the people of Namibia through a process of "Namibianization" of the fishing industry.

The Marine Resources Act, passed in 2000, required that all Namibian fisheries be managed on the basis of assigned fishing rights. These rights have been established as quota shares, valid for terms ranging from 7 to 20 years, depending on the fishery in question. A total of 163 such shares had been issued by 2002. In order to ensure Namibian ownership of, and benefits from the fishery, quota shares are non-transferable.

Monitoring, control and surveillance are major components of Namibia's fishery management system. First, major efforts have been made to prevent illegal fishing. In 1990–91, for example, 12 foreign vessels were arrested and successfully prosecuted for violating Namibia's EEZ. This action sent a strong signal to other potential poachers, and Namibia has far fewer problems in this area than many other developing countries. Second, all fishing vessels are licensed; by 2002 a total of 335 licenses were in effect, 71% of which were Namibian owned. Fishing activities are monitored by systematic sea and air patrols. A national satellite-based vessel monitoring system, similar to systems used elsewhere, is being implemented. Third, complete monitoring of all landings at the two commercial fishing ports of Walvis Bay and Lüderitz ensures that quota limits are upheld and fees paid. The fact that Namibia has no artisanal fisheries helps to simplify monitoring and control.

Fees are also an important part of the Namibian system. Quota fees earn substantial revenues for the government, while permitting profitable fishing for quota holders. By-catch fees attempt to deter the capture of non-target species, but are kept low enough not to encourage discarding, which is in fact prohibited. In view of the inevitability of some by-catch, a minor amount is allowed without penalty. On-board observers are present on most vessels. A separate fee is used to cover the costs of research, management and enforcement. These costs amounted to 3.6% of landed value in 1999.

Conservation measures include closed seasons to protect spawning, and closed areas to protect juvenile fish. The overall conservation-oriented management strategy has succeeded in rebuilding the stocks of Namibia's two largest fisheries, hake and horse mackerel; TACs of these species have increased from 60,000 (1990) to 195,000 tonnes (2002) and 150,000 (1990) to 350,000 tonnes (2002), respectively. A third important stock, pilchards, undergoes large-magnitude fluctuation, which necessitated a zero TAC in 2002 (20,000 tonnes in 2003).

The contribution of Namibia's fisheries to GDP increased from approximately N$ 300 million in 1991 (1%) to N$ 2,000 million in 2002 (6.6%). The fisheries sector is the second largest in the country, exceeded only by mining. Approximately 97% of landings are exported—for example, fresh hake fillets are transported regularly by air to European and other markets, in some cases providing products that are no longer available from depleted fisheries in other parts of the world. The fishing industry thus provides an important source of foreign revenues for the country.

Assessment

The Namibian government has put together a well-designed fisheries management system that appears to be generating rents that benefit the industry and the nation as a whole. Given the success of this system, perhaps a conscious program of risk management would be worth pursuing. To mention a single example, the possibility that bottom trawling may have long-term effects on demersal communities may be worth monitoring (see Kaiser et al. 2002). No doubt this and other forms of fine tuning the system will be employed to further enhance the contributions of Namibia's fisheries to the country's economy. The volume *Namibia's Fisheries* (Sumaila et al. 2004) provides further details. Bayesian analysis of parameter uncertainty for Namibian hake is discussed by Hilborn and Mangel (1997, p. 256).

6.5 British Columbia's Groundfish Trawl Fishery

A trawl fishery for groundfish off the coast of British Columbia (linear length approximately 900 km) has been in operation for over 60 years (Grafton et al. 2004). Dozens of species are caught, the most important being rockfish, hake, Pacific cod, thornyheads, sole and ling cod. Pacific halibut, another important species in the area, is taken in a separate hook-and-line fishery; any halibut taken by trawlers must be returned to the sea (even though mortality is near 100%).

Foreign fishing, prevalent in the 1960–1970s, was phased out following Canada's declaration of 200-mile fishing zones in 1977. Government subsidies on vessels and fish prices were used to encourage Canadian replacement of foreign effort. A total of 142 trawl vessels were licensed.

Management techniques included restrictions on vessels and gear, together with species TACs set for the entire coast. In addition, vessels were assigned non-transferable trip limits for each species. If a vessel's limit for any species was filled, the trip was required to terminate—in theory. In practice, the system lead to widespread discarding and misrepresenting of catches. The information provided to fisheries scientists became highly unreliable, forcing the managers to set progressively lower TACs which they knew would be exceeded to some unknown degree. Believing that control of the groundfishery had become lost, the government eventually closed the fishery entirely in September 1995.

The fishery was reopened in February 1996, using 100% on-board observers to record all catches. In 1997 the fishermen agreed to implement species ITQs on a trial basis, and this event signalled the beginning of a new approach to managing the trawl fishery.

After dividing the coastal TACs between trawl and hook-and-line sectors, the trawl quotas were further divided into a total of 55 separate species-area annual TACs, each of which was then allocated among the licensed vessels, as ITQ shares. The initial ITQ allocations were based on the size of the vessel and on its catch history. Twenty per cent of all ITQs were assigned to a new Groundfish Development Authority to be distributed with various aims, such as promoting regional development and employment and ensuring fair and safe treatment of crew.

The multispecies ITQ system works in the following manner. All catches, whether retained or discarded, are recorded and tallied against the vessel's ITQ by species. Once a vessel has caught its quota for any species in a given area, it must cease fishing in that area unless it obtains additional ITQ units. However, it is also possible to borrow from next year's quota, up to 37.5% of the vessel's permanent quota (15% for hake and halibut). Halibut by-catch quotas are part of the trawl ITQ system, but halibut can not be retained.

Over 2700 temporary, one-year transfers of quota units per year, among about 70 active trawlers, now take place. Average lease prices and permanent transfer prices for 2003/4 were 20¢ and $3.00 per "adjusted" lb, respectively. An adjusted lb (or Groundfish Equivalent) is the weight of a given species that has the same estimated landed value as one pound of Pacific Ocean perch.

(Officially, vessel owners do not own their ITQs on a permanent basis. Instead, the federal Minister of Fisheries grants the ITQs annually. In fact, with a few exceptions in the form of penalties, no ITQs have ever been withdrawn from their assigned vessels.)

Other rules pertaining to quota transfers set limits on quotas per vessel, and also regulate permanent transfers so as to prevent concentration of quota ownership, including ownership by inactive licensees. The transferability rules are revised every three years.

All trawlers are required to carry on-board observers while at sea. The cost of observers, approximately $300/day (2003/4), is paid by the vessel owners. To circumvent any risk of collusion, observers are assigned randomly to vessels each year.

The effects of the ITQ system have been profound. First, catch data are now reliable, allowing the scientists to perform believable TAC estimates. (This is the result of the observer program, not of the ITQ system itself, although the latter no doubt implies a degree of acceptance and support of the observer program.)

Second, a decrease in fleet capacity has occurred, as both small and large vessels have sold their quotas and withdrawn from the fishery. By 2003/4 about 70 vessels remained active in the trawl fishery. (For some reason, none of the original 142 vessel licenses have been retired, however.)

"The most significant change with the advent of ITQs has been in terms of fisher behavior." (Grafton et al. 2004, p. 11.) For example, fishers now tend to specialize in regions and species, in response to market opportunities. Shorter trips are used to improve the quality of the catch, for example, landing a higher proportion of fresh fish. In terms of resource conservation, discards are now not only accurately quantified, but have also been significantly reduced because of the ITQ-generated economic incentives against catching unwanted species. The existence of vessel quotas for discarded as well as retained species has provided fishermen with an incentive to be much more selective in their fishing behavior. Some species, such as Pacific Ocean perch and sablefish, are now seldom targeted directly because these species are caught incidentally in sufficient quantities, in conjunction with other species. Discard ratios (i.e., discards divided by retained catches) for many species have declined sharply since the ITQ system was initiated, indicating that fishermen have learned how to fish more efficiently.

With a total amount of catch of approximately 50 million lbs, and landed value about $35 million, the groundfish trawl fishery is one of the most important of Canada's West Coast fisheries.

Assessment

The most unusual feature of the B.C. groundfish trawl fishery under ITQs is its handling of multiple species and areas. One sometimes hears that ITQs are not appropriate for multispecies fisheries, owing to their complexity. This example strongly suggests that this may not be the case, provided that the ITQ system is meticulously (and adaptively) designed and enforced. As with other ITQ systems, the individual quotas alter the economic incentives of fishermen in a way that encourages both resource conservation and economically efficient harvesting.

Short-term transferability, or leasing, of quotas is clearly an essential feature of a multispecies ITQ system. Long-term permanent transferability is a separate issue, with its associated problems of quota concentration. The rule that quotas always be assigned to active fishing vessels (not exceeding realistic catch levels) would seem to ensure that concentration and outside ownership of quotas do not occur.

6.6 Overview

The five case studies presented here strongly suggest that ITQ-based (or related rights-based) management systems can affect fishermen's economic incentives in ways that encourage resource conservation and the preservation of economic benefits from a commercial fishery. These systems need to be carefully designed, preferably with the close cooperation of the fishing industry. In contrast to traditional approaches to fisheries management, which tended to pit managers versus harvesters, ITQ systems can involve close cooperation between fisheries managers and the industry.

The case studies also suggest that the difficulties of the transition from the old TAC-regulated system to a new rights-based system are by no means insurmountable. This transition, however, generates a new problem, that of "Whose fish?" or, more a propos, "Whose rents?" This question has been discussed in Sec. 4.8. Most of the case studies appear to show a timidity on the part of governments to face up to their responsibilities to the resource owners, the general public. Who will take the resource managers to court for this dereliction of duty? Time will tell.

For additional descriptions of ITQ systems see OSB (1999) and Hilborn et al. (2005). A detailed discussion of New Zealand's ITQ-based management system is given by Hersoug (2002).

7

Changing Direction

To a large extent the global crisis in marine fisheries has been a crisis in fisheries management. The typical institutional arrangement for managing fisheries has consisted of two main components (Figure 7.1), a scientific working group and a management, or decision-making body. The scientific group obtains data, either direct fishery data or independent survey data, or both, develops and fits models to the data, and formulates a set of recommendations for management. The management body then tries to reconcile these recommendations with the needs of the fishing industry, and determines the annual catch quota and other management regulations. In addition, an enforcement arm ensures that the quotas and regulations are obeyed. These three groups are established and financed by government, which may continue to play an overall supervisory role.

Two striking omissions from this description are any consideration of risk management (although this now appears to be changing), or of

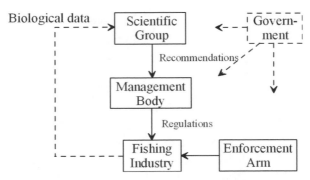

Figure 7.1 Typical institutional system for managing marine fisheries.

242

managing for profitability. Expertise in the latter areas is entirely lacking from many management structures.

7.1 A Responsible Management System

It could be argued with some strength that most governments have failed in their responsibility to protect valuable living marine assets within their 200-mile coastal zones. Some examples of inadequate management policy are:

1. Failure to identify and take account of major biological uncertainties.
2. Failure to recognize and take account of economic incentives that determine fishermen's behavior.
3. Failure to study and take account of the impacts of fishing gear (e.g., bottom trawls) on essential habitat.
4. Failure to learn from past mistakes. For example, when a managed fishery has collapsed, no concerted attempt may be made to discover exactly what went wrong (Smith and Link 2005).
5. Continued use of subsidies (including buy-back subsidies) in spite of growing evidence that these subsides often increase the likelihood of overfishing (Schrank and Keithly 1999; Munro and Sumaila 2002).
6. Failure to appreciate the scope and negative impact of illegal fishing (Sumaila et al. 2006).
7. Failure to recognize and take into account the public's right to receive a fair share, through resource royalties, of the net profits from the resource, which they own.

These failures all stem from the government's lack of recognition that fishery resource is a natural asset—a form of natural capital—owned by the nation, and in need of careful husbandry and management. The management system depicted in Fig. 7.1 is, as it stands, inadequate to this task.

At least two additional components are needed—first, a group of economics and decision-analysis experts, and second, a public overseer (Fig. 7.2). The public overseer holds the responsibility for policy formulation, and for ensuring that the fishery resource is managed for optimal long-term socio-economic benefits (sustainability, profitability, equity). The scientific and economic/decision-analysis groups work closely together to formulate recommendations for the management body. Operating under the rules set by the management body, the fishing industry

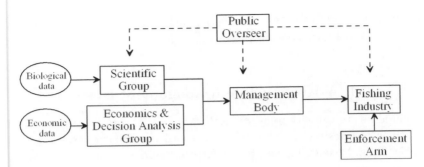

Figure 7.2 Revised institutional system for fisheries management.

generates optimal profits (net of management costs), which are shared equitably with government. In addition, the industry provides biological (catch, effort) data and economic (revenue, cost) data to the two working groups. Although the latter proviso might be considered outrageous under a traditional management system based on unallocated TACs, the provision of accurate economic data to analysts and managers would become an essential requirement under the new system.

Table 7.1 gives a brief comparison of the "Old" and "New" paradigms of fisheries management. The paradigms are obviously very different, and its seems clear that new institutional arrangements will be needed to operationalize the new paradigm.

Table 7.1. *Old and new paradigms.*

	Old	New
Resource owner	No one	The nation
Access	Free and unrestricted	Limited to quota owners
Management objective	Maximum sustainable yield	Sustainability and profitability
The fishing "game"	Non-cooperative	Cooperative
Risk management	Not usually considered	Assumed to be essential
Equilibrium	Regulated bio-nomic equilibrium	A sustainable, profitable fishery
Role of the ecosystem	Not considered	Recognized and taken into account

The overseer body will set its own criteria for management. Here is a checklist of topics that ought to be considered:

1. What system boundaries?
   - single species, community ecology, or ecosystem?
   - allow for needs of predators, e.g., birds, mammals?
2. What management objectives?
   - profitability, sustainability, equity
   - constrained optimization
3. What basic management system?
   - TACs only, IFQs or ITQs, community rights?
4. What data?
   - CPUE data, survey data, meta-data?
5. How will uncertainty be treated?
   - identify major sources of uncertainty
   - use decision (policy) analysis
6. Which model?
   - single species, multi-species, ecosystem?
   - spatially structured?
7. What risk-management strategies?
   - control the risk of overfishing or collapse
   - diversification, marine reserves (closed areas)?
8. What monitoring and enforcement system?
   - on-board observers, satellite transponders?
9. How to evaluate the management system?
   - independent reviews?
   - "what-might-go-wrong" approach?

This list is incomplete, but it does contain several novel components that are consistent with the methods of decision analysis.

High-Seas Fisheries

Although this book does not deal explicitly with high-sea fisheries (i.e. fisheries that occur beyond any 200-mile EEZ), the possibility does exist that these fisheries may eventually be placed under circumstances in which the principles of management for sustainable, profitable fishing could apply. For example, it might be agreed under a United Nations umbrella that certain oceanic zones would become joint EEZs, controlled jointly by specified regional states. These states would then set up a joint management system, with allocated (and perhaps transferable) quotas to each participating state. International equity would imply profit sharing between quota owners and the international community.

This description may seem a fantasy, but without some such system of jurisdiction, the present crisis of devastation of high-sea fish resources can only grow worse.

Appendix

This Appendix briefly considers some aspects of fisheries management that have not been fully discussed in the book. References are cited for additional information, but few of these topics have yet been studied in any detail, certainly not in a bioeconomic setting.

Stock Assessment

Techniques of stock assessment began changing rapidly around 1995 (see Hilborn et al. 1993; Punt and Hilborn 1997; McAllister and Kirkwood 1998; Parma 2001; Walters and Martell 2002; Quinn and Collie 2005). Contemporary stock assessment uses Bayesian methods (which are computationally intensive) to quantify the level of uncertainty in these estimates (see Chapter 5).

The opinion is sometimes encountered that, having failed in the past to forecast population collapses, stock assessment is no longer needed in fisheries management. I believe that this is a dangerous trend. Managers need to have up-to-date knowledge of the state of the stocks, as well as a realistic assessment of the degree of uncertainty pertaining to this information. The question of how uncertainties in stock assessment (and elsewhere) should be addressed by the decision makers was discussed in Chapters 5 and 7.

Simulation Models

The models used in this book are analytic models, that is, models that are written down in mathematical language, or specified graphically. Although analytic models are useful for obtaining insights about fisheries management, they are usually not appropriate for specific use, for example in setting or allocating actual annual quotas. The advantages of analytic models are simplicity, generality, transparency, and relative ease of solution or analysis. The main disadvantage is the limited complexity that can be included in such models.

Computer simulation models, on the other hand, can encompass almost unlimited complexity, tailored to the specific fishery being studied. A disadvantage of simulation models, especially those that employ a pre-existing software package, is that important model details may be concealed from the user. Thus the reasons underlying model predictions may be obscure. If these predictions seem counter-intuitive, no insights may arise from the simulation exercise. The solution to this dilemma is for modelers to write their own simulation codes. In some cases a spreadsheet will be sufficient for this purpose. In any case, the increasing use of simulation models, especially in decision analysis, seems certain to occur.

Ecosystem Models

It is currently controversial whether fishery management based on single-species models should now be replaced by ecosystem-based management, using complex ecosystem models.

While there can be little doubt that overfishing has often had severe impacts on marine ecosystems (Jackson et al. 2001), it is not yet clear whether ecosystem modeling will help much, at least in a quantitative sense. The problems noted above for simulation models will only intensify with the level of complexity required to model full ecosystems. In addition, the adequacy of data to support such models is questionable. Perhaps the main role for ecosystem models will be as research tools, used to investigate new management strategies that might reduce the risk of system collapse or irreversible change.

Social Cost and Unemployment

Coastal communities in many countries suffer from high levels of unemployment. If workers who would otherwise be unemployed (more or less permanently) are hired as crew or plant workers, a social cost–benefit analysis of the fishery should not include the wages paid to these individuals as part of the operating cost of fishing or of processing the catch. In this case the social cost of labor is smaller than the actual wages paid to crew by the vessel owner or plant operator. This in turn means that wage subsidization may be socially justified (Clark and Munro 1975).

There should be no need, however, to use public funds directly to subsidize the wages of workers. Instead, part of the profits (rents) from operating a well-managed fishery would constitute transfer payments from the vessel owners to the workers. Catch royalties should then be reduce accordingly, to allow for these social transfers.

Similar considerations apply to the case of artisanal fisheries. This term applies to traditional fisheries that have long been a way of life in coastal communities. Artisanal fisheries can be devastated when a large-scale "industrial" fishery begins exploiting offshore stocks. To mention one example, in Canada the annual catches of Atlantic cod by small dories fishing close to shore were sharply reduced when offshore trawlers began exploiting offshore segments of the cod population. Such offshore, large-scale fisheries may appear "efficient" from a cost–benefit stand-point, but if the inshore fishermen are displaced or thrown out of work, their loss of income needs to be treated as a social cost. See Charles (2001) for further discussion.

Economic Efficiency

The term "economic efficiency" can be used in two different ways. For an individual vessel, efficiency sometimes refers to the effort capacity of the vessel. Various capital improvements, such as improved naviga-tional equipment or more powerful winches, can increase the vessel's efficiency in this sense. Under regulated open access (or under individ-ual effort quotas), vessel owners are motivated to increase efficiency by such methods, i.e., by capital stuffing. As noted in Chapter 3, these in-creases in vessel efficiency are often counterproductive for the fishery as a whole.

A second meaning of economic efficiency pertains to cost efficiency, i.e. the cost of catching fish. Under an IFQ system, fishermen will be motivated to maximize cost efficiency, by minimizing the total cost of catching their annual quotas. Such increases in efficiency are in fact beneficial in an overall economic sense.

Fishing Capacity

Chapter 3 discussed questions of overcapacity and optimal capacity. The term "capacity" was conceptualized as effort capacity, denoted by $E_{\max}$, in the sense that

$$E(t) \leq E_{\max}$$

It was also assumed that fixed capital costs for the whole fleet are pro-portional to $E_{\max}$:

$$\text{fixed capital cost} = c_f E_{\max}$$

This means that fleet capacity is identified as the number of standardized vessels currently available for use in the fishery. This seems like a good first approximation to the concept of fishing capacity.

Economists have developed more complex models of fishing capacity, however (Pascoe and Gréboval 2003). For example, the capacity concept can be extended to include both effort capacity and catching capacity. The latter is related to such vessel characteristics as hold or freezer capacity.

If H_{max} denotes catch capacity (tonnes/day), then the extended view of fishing capacity implies an additional constraint

$$H(t) \leq H_{max}$$

For example, the derby fishery model discussed in Chapters 2 and 4 could be modified in this way.

The discussion in Chapter 3 defined "excess capacity" as any amount of fleet effort capacity greater than that required to capture the annual catch quota, or TAC. This definition is problematic, however, in the event that the TAC varies randomly from year to year. How should optimal capacity (and hence overcapacity) be defined in this case? Should it be capable of just taking the average TAC, or the largest likely TAC, or what? And what would an ITQ system imply for fleet capacity in such situations? It should be possible to extend the bioeconomic models of Chapters 3 and 4 to deal with these questions, but this research has not been carried out as yet.

Transboundary Fisheries

Many fish populations cross international boundaries, either because of their annual migration patterns, or through the dispersal of larvae or young fish. In such cases, the fishing activities of one state may influence the availability of fish for the neighboring state. International joint commissions have sometimes been established to control the activities of both states (or more accurately, to recommend such controls) for the benefit of each. Typically the commission is charged with performing stock assessments and recommending overall TACs (and perhaps seasonal openings), leaving the allocation of the TAC to negotiations between the states. In the absence of such agreement and negotiation, the danger exists that he states may participate in a free-for-all, with undesirable results. See Munro (1979; 1990) for game-theoretic models of transboundary fisheries.

Management Costs

The costs of managing a fishery can be substantial. For example, the cost of research (stock surveys, life history studies, measurement of environmental and ecosystem effects, etc.) may be large and on-going. Monitoring and enforcement of regulations may also be costly. Evidence from existing IFQ, and especially ITQ, systems, however, suggests that cooperation by the fishing industry may help reduce such management costs. ITQ holders favor, and may support, research that enhances the value of their quotas. Likewise, they may cooperate in ensuring that quotas are adhered to, and other regulations upheld.

Ocean Pollution

The oceans are the ultimate destination of many pollutants that originate on land. Examples include agricultural runoff, outflow from polluted rivers, and atmospheric aerosols that eventually fall or diffuse into the sea. Also, oil spills and ocean dumping of wastes add pollutants directly to the oceans. The biological effects of these pollutants, though largely unknown, are likely to be negative in terms of productivity of marine populations (Sindermann 1995). For example, Woodward (2000) reports that virtually all marine life other than jellyfish has disappeared from the Black Sea, because of hypoxia induced by discharges from the Danube, Dneister, Dneiper and Don Rivers. Hypoxic conditions are now chronic in many coastal seas and bays, including the Gulf of Mexico, the Adriatic, North and Baltic Seas, and the Sea of Japan and the Yellow Sea, among others. In addition to the loss of fishery resources, the costs of marine pollution include the collapse of coastal fishing communities and the loss of tourist revenues.

It has recently been noted that acidity levels in the ocean are rising as a result of increased concentrations of atmospheric carbon dioxide (Royal Society 2005). This process, which is irreversible over tens of thousands of years, could eventually impact the growth and survival of coral reefs and plankton communities.

The economics of ocean pollution resemble that of atmospheric pollution, involving externalities, common-pool resources, and future discounting. In particular, the lack of individual (or even national) ownership of ocean areas implies the lack of any individual incentive to protect these resources. Conceivably, however, allocated pollution quotas for toxic or otherwise damaging discharges, akin to present-day tradeable atmospheric pollution permits might be introduced at some future time. If so, it is perhaps also conceivable that the "owners" of fishery resources—the ITQ owners—would purchase some of the pollution per-mits with the aim of retiring them, thereby reducing ocean pollution. Other than this utopian outcome, the control of ocean pollution will depend on government regulation and international agreements. The current stalemate on the Kyoto accord does not bode well for the future control of ocean pollution, however.

Bibliography

Apostolaki, P., Milner-Gulland, E.J., McAllister, M.K. and Kirkwood, G.P. (2002). Modelling the effects of establishing a marine reserve for mobile fish species, *Canadian Journal of Fisheries and Aquatic Sciences* **59**, 405–415.

Arnason, R. (1993). The Icelandic individual transferable quota system: a descriptive account, *Marine Resource Economics* **8**, 201–218.

Arrow, K.J. (1968). Optimal capital policy with irreversible investment, in *Value, Capital and Growth: Papers in Honor of Sir John Hicks*, ed. J.N. Wolfe (Edinburgh University Press, Edinburgh, U.K.).

Barrett, S. (2003). *Environment and Statecraft: The Strategy of Environmental Treaty-Making*, (Oxford University Press, Oxford, U.K.).

Beddington, J.R. and May, R.M. (1977). Harvesting natural populations in a fluctuating environment, *Science* **197**, 463–465.

Berck, P. and Perloff, J.M. (1984). An open-access fishery with rational expectations, *Econometrica* **52**, 489–506.

Berkeley, S.A., Hixon, M.A., Larson, R.J. and Love, M.S. (2004). Fisheries sustainability via protection of age structure and spatial distribution of fish populations, *Fisheries* (Aug. 2004), 23–31.

Beverton, R.J.H. and Holt, S.J. (1957). On the Dynamics of Exploited Fish Populations, *Fisheries Investigation Series* 2(19), (Ministry of Agriculture, Fisheries and Food, London, U.K.).

Bjørndal, T., Ussif, A. and Sumaila, U.R. (2004). A bioeconomic analysis of the Norwegian spring spawning herring stock, *Marine Resource Economics* **19**, 353–365.

Caddy, J.F. and Seijo, J.C. (2005). This is more difficult than we thought! The responsibility of scientists, managers and stakeholders to mitigate the unsustainability of marine fisheries, *Philosophical Transactions of the Royal Society B* **360**, 59–75.

Charles, A.T. (2001). *Sustainable Fishery Systems*, (Blackwell Science, Oxford, U.K.).

Charles, A.T., Mazany, R.L. and Cross, M.L. (1999). The economics of illegal fishing: a behavioral model, *Marine Resource Economics* **19**, 95–110.

Christy, F.T. Jr. and Scott, A.D. (1965). *The Common Wealth in Ocean Fisheries*, (Johns Hopkins University Press, Baltimore, MD).

Ciriacy-Wantrup, S.V. (1972). *Resource Conservation: Economics and Policy.* 2nd edition, (University of California Press, Berkeley, CA).

Clark, C.W. (1990). *Mathematical Bioeconomics: The Optimal Management of Renewable Resources*, 2nd edition, (John Wiley and Sons, New York, NY); Paperback Edition 2005.

Clark, C.W. and Lamberson, R.H. (1982). An economic history and analysis of pelagic whaling, *Marine Policy* **6**, 103–120.

Clark, C.W. and Munro, G.R. (1975). The economics of fishing and modern capital theory: a simplified approach, *Journal of Environmental Economics and Management* **2**, 92–106.

Clark, C.W., Clarke, F.H. and Munro, G.R. (1979). The optimal exploitation of renewable resource stocks: problems of irreversible investment, *Econometrica* **47**, 25–41.

Clark, C.W., Munro, G.R. and Sumaila, U.R. (2005). Subsidies, buybacks and sustainable fisheries, *Journal of Environmental Economics and Management* **50**, 47–58.

Clarke, F.H. and Munro, G.R. (1987). Coastal states, distant water fishing nations and extended jurisdiction: a principal-agent analysis, *Natural Resource Modeling* **2**, 81–107.

Cooke, J.G. and Beddington, J.R. (1985). The relationship between catch rates and abundance in fisheries, *Journal of Mathematical Biology* **1**, 391–406.

Crowder, L., Lyman, S., Figueira, W. and Priddy, J. (2000). Source-sink population dynamics and the siting of marine reserves, *Bulletin of Marine Science* **66**, 799–820.

Dugatkin, L.A. (1997). *Cooperation among animals: an evolutionary perspective*, (Oxford University Press, New York, NY).

Dulvy, N.K., Sadovy, Y. and Reynolds, J.D. (2003). Extinction vulnerability in marine populations, *Fish and Fisheries* **4**, 24–64.

Edwards, S.F. (2002). Rent seeking and property rights formation in the U.S. Atlantic sea scallop fishery, *Marine Resource Economics* **16**, 263–275.

Eggertson, T. (2004). The subtle art of major institutional change: introducing property rights in the Iceland fisheries, in *Role of Institutions in Rural Policies and Agricultural Models*, eds. G. van Huyleborook, W. Verkeke and L. Lauwers, (Elsevier, Netherlands) pp. 43–59.

Fox, W.W. (1970). An exponential surplus yield model for optimizing in exploited fish populations, *Transactions of the American Fisheries Society* **99**, 80–88.

Francis, R.I.C.C. and Schotten, R. (1997). "Risk" in fisheries management: a review, *Canadian Journal of Fisheries and Aquatic Sciences* **54**, 1699–1715.

Gordon, H.S. (1954). The economic theory of a common property resource: the fishery, *Journal of Political Economy* **62**, 124–142.

Grafton, R.Q., Nelson. H.W. and Turris, B. (2004). How to resolve the Class II common property problem? The case of British Columbia's multi-species groundfish trawl fishery. Draft.

Grafton, R.Q. and 18 other authors. (2005). Incentive-based approach to fisheries management. *Canadian Journal of Fisheries and Aquatic Sciences* (In press).

Gréboval, D. and Munro, G.R. (1999). Overcapitalization and excess capacity in world fisheries: underlying economics and methods of control, in *Managing Fishing Capacity*, ed. D. Gréboval (Food and Agriculture Organization of the United Nations, FAO Technical Paper 386 (Rome), pp. 1–48.

Gulland, J.A. (1964). The reliability of catch per unit effort as a measure of abundance of the North Sea trawl fisheries, in *On the Measurement of Abundance of Fish Stocks*, ed. J.A. Gulland, *J. du Conseil Intern. Pour l'Expl. de la Mer, Rapports de Proc. Verbaux des Reunions* **177**, 98–102.

Gulland, J.A. (1983). *Fish Stock Assessment*, (Wiley, Chichester, UK).

Hannesson, R. (1998). Marine reserves: what would they accomplish? *Marine Resource Economics* **13**, 159–170.

Harwood, J. and Stokes, K. (2003). Coping with uncertainty in ecological advice: lessons from fisheries, *Trends in Ecology and Evolution* **18**, 617–622.

Hastings, A. and Botsford, L.W. (1999). Equivalence in yield from marine reserves and traditional fisheries management, *Science* **284**, 1537.

Hersoug, B. (2002). *Unfinished Business: New Zealand's Experience with Rights-based Fisheries Management*, (Eburon, Delft).

Hilborn, R. (2002). The dark side of reference points, *Bulletin of Marine Science* **70**, 403–408.

Hilborn, R., Branch, T.A., Ernst, B., Magnusson, A., Minte-Vera, C.V., Scheurell, M.D. and Valero, J.L. (2003). State of the World's Fisheries, *Annual Rev. Environmental Resources* **28**, 359–399.

Hilborn, R. and Mangel, M. (1997). *The Ecological Detective: Confronting Models with Data*, (Princeton University Press, Princeton, NJ).

Hilborn, R., Orensanz, J.M. and Parma. A.M. (2005). Institutions, incentives and the future of fisheries, *Philosophical Transactions of the Royal Society B* **360**, 47–57.

Hilborn, R., Pikitch, E.K. and Francis, R.C. (1993). Current trends in including risk and uncertainty in stock assessment and harvest decisions. *Canadian Journal of Fisheries and Aquatic Sciences* **50**, 874–880.

Hilborn, R. and Walters, C.J. (1992). *Quantitative Fisheries Stock Assessment: Choice, Dynamics and Uncertainty*, (Chapman and Hall, New York, NY).

Holland, D.S. (2004). Spatial fishery rights and marine zoning: a discussion with reference to management of marine resources in New England, *Marine Resource Economics* **19**, 21–40.

Holland, D., Gudmundsson, E. and Gates, J. (1999). Do fishing vessel buyback programs work? A survey of the evidence, *Marine Policy* **23**, 47–69.

Holland, D.S. and Brazee, R.J. (1996). Marine reserves for fisheries management, *Marine Resource Economics* **3**, 157–171.

Homans, F.R. and Wilen, J.E. (1997). A model of regulated open access resource use, *J. Environmental and Resource Economics* **32**, 1–21.

Hotelling, H. (1931). The economics of exhaustible resources, *Journal of Political Economy* **39**, 137–175.

Huppert, D.D. (2006). Partial auctions of IFQs: A means to share the "rent." (in press)

Hutchings, J.A. (2000). Collapse and recovery of marine fishes, *Nature* **406**, 882–885.

Jackson, J.B.C. and 18 other authors. (2001). Historical overfishing and the recent collapse of coastal ecosystems, *Science* **293**, 629–638.

James, M. (2004). The British Columbia salmon fishery "buyback" program—a case study in capacity reduction, (IIFET Japan Conference Proceedings) Unpublished.

Jannson, A.-M., Hammer, M., Folke, C. and Costanza, R. (1994). *Investing in Natural Capital: The Ecological Economics Approach to Sustainability*, (Island Press, Washington, D.C.).

Kaiser, M.J., Collie, J.S., Hall, S.J., Jennings, S. and Poiner, I.R. (2002). Modification of marine habitats by trawling activities: prognosis and solutions, *Fish and Fisheries* **3**, 114–136.

Kuperan, K. and Sutinen, J.G. (1998). Blue-water crime: Deterrence, legitimacy, and compliance in fisheries, *Law and Society Review* **32**, 309–338.

Lane, D.E. and Stephenson, R.L. (1998). A framework for risk analysis in decision-making, *ICES Journal of Marine Science* **55**, 1–13.

Larkin, P. (1977). An epitaph for the concept of maximum sustained yield, *Trans. Amer. Fisheries Soc.* **106**, 1–11.

Lauck, T., Clark, C.W., Mangel, M., and Munro, G.R. (1998). Implementing the precautionary principle in fisheries through marine reserves, *Ecological Applications* **8** (Suppl.), S72–S80.

Liermann, M. and Hilborn, R. (2001). Depensation: evidence, models and implications, *Fish and Fisheries* **2**, 33–58.

Macinko, S. and Bromley, D.W. (2002). *Who owns America's fisheries?* (Center for Resource Economics, Covelo, CA).

Mangel, M. and 42 other authors. (1996). Principles for the conservation of wild living resources, *Ecological Applications* **6**, 338–362.

Mangel, M. and Levin, P.S. (2005). Regime, phase and paradigm shifts: making community ecology the basic science for fisheries management, *Philosophical Transactions of the Royal Society B* **360**, 95–105.

Mangel, M., Marinovic, B., Pomeroy, C. and Croll, D. (2002). Requiem for Ricker: unpacking MSY, *Bulletin of Marine Science* **70**, 763–781.

Matulich, S.C. and Clark, M.L. (2003). North Pacific halibut and sablefish IFQ policy design: Quantifying the impacts on processors. *Marine Resource Economics* **18**, 149–166.

May, R.M., Beddington, J.R., Clark, C.W., Holt, S.J. and Laws, R.M. (1979). Management of multispecies fisheries, *Science* **205**, 267–277.

Maynard Smith, J. (1982). *Evolution and the Theory of Games*, (Cambridge University Press, Cambridge, U.K.).

McAllister, M. and Kirchner, C. (2002). Accounting for structural uncertainty to facilitate precautionary fishery management: illustration with Namibian orange roughy, *Bulletin of Marine Science* **70**, 499–540.

McAllister, M. and Kirkwood, G.P. (1998). Bayesian stock assessment: a review and example application using the logistic model, *Journal of Marine Science* **55**, 1031–1060.

McKelvey, R. (1985). Decentralized regulation of a common property resource industry with irreversible investment, *Journal of Environmental Economics and Management* **12**, 287–307.

McKelvey, R. (1986). Fur seal and blue whale: the bioeconomics of extinction, in *Applications of Control Theory in Ecology*, ed Y. Cohen. *Lecture Notes in Biomathematics* **73** (Springer-Verlag, Berlin) pp. 57–82.

McRae, D. and Munro, G.R. (1989). Coastal state "rights" within the 200-mile exclusive economic zone, in *Rights Based Fishing*, ed. P.A. Neher et al. (Kluwer, Amsterdam), pp. 97–111.

Mesterton-Gibbons, M. and Adams, E.S. (2002). The economics of animal cooperation, *Science* **298**, 2146–2147.

Milon, J.W., Larkin, S.L. and Eberhardt, N.M. (1999). Bioeconomic models of the Florida commercial spiny lobster fishery, *Sea Grant Report* **117** (University of Florida, Gainesville, FL).

Morgan, M.G. and Henrion, M. (1990). *Uncertainty: A guide to treatment of uncertainty in quantitative policy analysis*, (Cambridge University Press, Cambridge, U.K.).

Mullon, C., Freon, P. and Cury, P. (2005). The dynamics of collapse in world fisheries, *Fish and Fisheries* **6**, 111–120.

Munro, G.R. (1979). The optimal management of transboundary renewable resources, *Canadian Journal of Economics* **12**, 355–376.

Munro, G.R. (1987). The management of shared fishery resources under extended jurisdiction, *Marine Resource Economics* **3**, 271–296.

Munro, G.R. (1990). The optimal management of transboundary fisheries: game-theoretic considerations, *Natural Resource Modeling* **4**, 403–426.

Munro, G.R. and Sumaila, U.R. (2002). Subsidies and their potential impact on the management of the ecosystems of the North Atlantic, *Fish and Fisheries* **3**, 233–250.

Munro, G.R., Van Houtte, A. and Willmann, R. (2004), The Conservation and Management of Shared Fish Stocks: Legal and Economic Aspects, *FAO Fisheries Technical Paper* **465** (Rome).

Murawski, S.A., Brown, R., Kai, H.-L., Rago, P.J. and Hendrickson, J. (2000). Large-scale closed areas as a fisheries management tool in temperate marine systems: the Georges Bank experience, *Bulletin of Marine Science* **66**, 775–798.

Myers, R.A., Hutchings, J.A. and Borrowman, N.I. (1997). Why do fish stocks collapse? The example of cod in Atlantic Canada, *Ecological Applications* **7**, 91–106.

Myers, R.A. and Mertz, G. (1998). The limits of exploitation: a precautionary approach, *Ecological Applications* **8** (Suppl.), S165–S169.

Myers, R.A. and Worm, B. (2003). Rapid worldwide depletion of predatory fish communities, *Nature* **423**, 280–283.

Nash, J.F. (1951). Noncooperative games, *Annals of Mathematics* **54**, 289–295.

Nash, J.F. (1953). Two-person cooperative games, *Econometrica* **21**, 128–140.

Neher, P.A. (ed.). *Rights Based Fishing*, (Kluwer, Amsterdam).

Nichols, P. (2004). Marine fisheries management in Namibia: Has it worked? in *Namibia's Fisheries: Ecological, Economic and Social Apsects*, ed. U.R. Sumaila, D. Boyer, M.D. Skogen and S.I. Steinsham (Eburon, Delft), pp. 319–332.

Ocean Studies Board of the USA. (1999). *Sharing the Fish: Toward a National Policy on Individual Fishing Quotas*, (National Academy Press, Washington, DC).

Ostrom, E. (1990). *Governing the Commons: The Evolution of Institutions for Collective Action*, (Cambridge University Press, Cambridge, U.K.).

Parma, A.M. (2001). Bayesian approaches to the analysis of uncertainty in the stock assessment of Pacific halibut, in *Incorporating Uncertainty into Fishery Models*, ed. J.M. Berkson, L.L. Kline and D.J. Orth (American Fisheries Society, Bethesda, MD).

Pascoe, S. and Gréboval, D. eds. (2003). *Measuring Capacity in Fisheries*, (Food and Agriculture Organization of the United Nations, Rome), Technical Paper 445.

Pauly, D. (1995). Anecdotes and the shifting baseline syndrome of fisheries, *Trends in Ecology and Evolution* **10**, 430.

Pauly, D., Christensen, V., Dalsgaard, J., Froese, R. and Torres, F. (1998). Fishing down marine food webs, *Science* **279**, 860–863.

Pauly, D. and Maclean, J. (2003). *In a Perfect Ocean: The State of Fisheries and Ecosystems in the North Atlantic Ocean*, (Island Press, Washington, DC).

Pikitch, E.K. and 16 other authors. (2004). Ecosystem-based fishery management, *Science* **305**, 346–347.

Punt, A.E. and Hilborn, R. (1997). Fisheries stock assessment and decision analysis: A review of the Bayesian approach, *Rev. Fish Biology Fisheries* **7**, 35–63.

Quinn, T.J. (2003). Rumination on the development and future of population models in fisheries, *Natural Resource Modeling* **16**, 341–392.

Quinn, T.J. II and Collie, J.S. (2005). Sustainability in single-species population models, *Philosophical Transactions of the Royal Society B* **360**, 147–162.

Quinn, T.J. and Deriso, R.B. (1999). *Quantitative Fish Dynamics*, (Oxford University Press, New York, NY).

Repetto, R. (2001). A natural experiment in fisheries management, *Marine Policy* **25**, 251–264.

Ricker, W.E. (1954). Stock and Recruitment, *J. Fish. Res. Board Canada* **11**, 559–623.

Ridley, M. (1998). *The Origins of Virtue: Human Instincts and the Evolution of Cooperation*, (Penguin Books, New York, NY).

Roberts, C.M., Hawkins, J.P. and Gell, F.R. (2005). The role of marine reserves in achieving sustainable fisheries, *Philosophical Transactions of the Royal Society B* **360**, 123–132.

Rosenberg, A.A. and Restrepo, V.R. (1994). Uncertainty and risk evaluation in stock assessment advice for U.S. marine fisheries, *Canadian Journal of*

Fisheries and Aquatic Sciences **51**, 2715–2720.

Rothschild, B.J. (1972). An exposition on the definition of fishing effort, *Fishery Bulletin* **70**, 671–679.

Royal Society. (2005). Ocean acidification due to increasing atmospheric carbon dioxide, Royal Society Policy Doc. 12/05.

Sanchirico, J.N. and Wilen, J.E. (1999). Bioeconomics of spatial exploitation in a patchy environment, *Journal of Environmental Economics and Management* **37**, 129–150.

Sanchirico, J.N. and Wilen, J.E. (2001). A bioeconomic model of marine reserve creation, *Journal of Environmental Economics and Management* **42**, 257–276.

Sanchirico, J.N. and Wilen, J.E. (2005). Optimal spatial management of renewable resources: matching policy scope to ecosystem scale, *Journal of Environmental Economics and Management* **50**, 23–46.

Schaefer, M.B. (1954). Some aspects of the dynamics of populations important to the management of commercial fisheries, *Bulletin of the Inter-American Tropical Tuna Commission* **1**, 25–56.

Schrank, W.E. and Keithly Jr., W.R. (1999). The concept of subsidies, *Marine Resource Economics* **14**, 151–164.

Scott, A.D. (1955). The fishery: the objectives of sole ownership, *Journal of Political Economy* **63**, 116–124.

Sindermann, C.J. (1995). *Ocean Pollution*, (C&C Press, Boca Raton, FL).

Smith, T.D. (1994). *Scaling Fisheries*, (Cambridge University Press, Cambridge, U.K.).

Smith, T.D. and Link, J.L. (2005). Autopsy your dead. . . and living: a proposal for fisheries science, fisheries management and fisheries, *Fish and Fisheries* **6**, 73–87.

Smith, V.L. (1969). On models of commercial fishing, *Journal of Political Economy* **77**, 181–198.

Spagnolo, M. (2004). Property rights as a management tool in the Mediterranean: the Atlantic clam fishery. (Draft).

Stephens, D.W., McLinn, C.M. and Stevens, J.R.. (2002). Discounting and reciprocity in an iterated prisoner's dilemma, *Science* 298, 2216–2218.

Sumaila, U.R. and Alder, J. Eds. (2001). *Economics of Marine Protected Areas*, (Fisheries Centre, University of British Columbia, Vancouver, B.C.).

Sumaila, U.R., Alder, J. and Keith, H. (2006). Global scope and economics of illegal fishing, *Marine Policy* (in press).

Sumaila, U.R. and Walters, C.J. (2005). Intergenerational discounting: a new intuitive approach, *Ecological Economics* **52**, 135–142.

Sumaila, U.R., Boyer, D., Skogen, M.D. and Steinsham, S.I. Eds. (2004). *Namibia's Fisheries: Ecological, Economic and Social Aspects*, (Eburon, Delft).

von Neumann, J. and Morgenstern, O. (1947). *The Theory of Games and Economic Behavior*, (Princeton University Press, Princeton, NJ).

Walters, C.J. (1986). *Adaptive Management of Renewable Resources*, (MacMillan Publishing Company, New York, NY).

Walters, C., Christensen, V. and Pauly, D. (1997). Structuring dynamic models of exploited ecosystems from trophic mass-balance assessments, *Reviews Fish Biology and Fisheries* **7**, 1–34.

Walters, C.J. and Martell, S.J.D. (2002). Stock assessment needs for sustainable fisheries management, *Bulletin of Marine Science* **70**, 629–638.

Walters, C.J. and Martell, S.J.D. (2004). *Fisheries Ecology and Management*, (Princeton University Press, Princeton, NJ).

Weninger, Q. and McConnell, K.E. (2000). Buyback programs in commercial fisheries: efficiency versus transfers, *Canadian Journal of Economics* **33**, 394–412.

Weitzman, M.L. (2002). Landing fees vs. harvest quotas with uncertain fish stocks, *Journal of Environmental Economics and Management* **43**, 325–338.

Wilen, J.E. (2004). Spatial management of fisheries, *Marine Resource Economics* **19**, 7–19.

Woodward, C. (2000). *Ocean's End*, (Basic Books, New York, N.Y.).

Yamamoto, T. (1995). Development of a community-based fishery management system in Japan, *Marine Resource Economics* **10**, 21–34.

Index